Peer-to-Peer Data Management

Synthesis Lectures on Data Management

Editor
M. Tamer Özsu, ***University of Waterloo***

Synthesis Lectures on Data Management is edited by Tamer Özsu of the University of Waterloo. The series will publish 50- to 125 page publications on topics pertaining to data management. The scope will largely follow the purview of premier information and computer science conferences, such as ACM SIGMOD, VLDB, ICDE, PODS, ICDT, and ACM KDD. Potential topics include, but not are limited to: query languages, database system architectures, transaction management, data warehousing, XML and databases, data stream systems, wide scale data distribution, multimedia data management, data mining, and related subjects.

Peer-to-Peer Data Management
Karl Aberer
2011

Probabilistic Ranking Techniques in Relational Databases
Ihab F. Ilyas and Mohamed A. Soliman
2011

Uncertain Schema Matching
Avigdor Gal
2011

Fundamentals of Object Databases: Object-Oriented and Object-Relational Design
Suzanne W. Dietrich and Susan D. Urban
2010

Advanced Metasearch Engine Technology
Weiyi Meng and Clement T. Yu
2010

Web Page Recommendation Models: Theory and Algorithms
Sule Gündüz-Ögüdücü
2010

Multidimensional Databases and Data Warehousing
Christian S. Jensen, Torben Bach Pedersen, and Christian Thomsen
2010

Database Replication
Bettina Kemme, Ricardo Jimenez Peris, and Marta Patino-Martinez
2010

Relational and XML Data Exchange
Marcelo Arenas, Pablo Barcelo, Leonid Libkin, and Filip Murlak
2010

User-Centered Data Management
Tiziana Catarci, Alan Dix, Stephen Kimani, and Giuseppe Santucci
2010

Data Stream Management
Lukasz Golab and M. Tamer Özsu
2010

Access Control in Data Management Systems
Elena Ferrari
2010

An Introduction to Duplicate Detection
Felix Naumann and Melanie Herschel
2010

Privacy-Preserving Data Publishing: An Overview
Raymond Chi-Wing Wong and Ada Wai-Chee Fu
2010

Keyword Search in Databases
Jeffrey Xu Yu, Lu Qin, and Lijun Chang
2009

Peer-to-Peer Data Management

Karl Aberer

www.morganclaypool.com

ISBN: 9781608457199 paperback
ISBN: 9781608457205 ebook

DOI 10.2200/S00338ED1V01Y201104DTM015

A Publication in the Morgan & Claypool Publishers series
SYNTHESIS LECTURES ON DATA MANAGEMENT

Lecture #15
Series Editor: M. Tamer Özsu, *University of Waterloo*
Series ISSN
Synthesis Lectures on Data Management
Print 2153-5418 Electronic 2153-5426

Peer-to-Peer Data Management

Karl Aberer
EPFL

SYNTHESIS LECTURES ON DATA MANAGEMENT #15

ABSTRACT

This lecture introduces systematically into the problem of managing large data collections in peer-to-peer systems. Search over large datasets has always been a key problem in peer-to-peer systems and the peer-to-peer paradigm has incited novel directions in the field of data management. This resulted in many novel peer-to-peer data management concepts and algorithms, for supporting data management tasks in a wider sense, including data integration, document management and text retrieval.

The lecture covers four different types of peer-to-peer data management systems that are characterized by the type of data they manage and the search capabilities they support. The first type are *structured peer-to-peer data management systems* which support structured query capabilities for standard data models. The second type are *peer-to-peer data integration systems* for querying of heterogeneous databases without requiring a common global schema. The third type are *peer-to-peer document retrieval systems* that enable document search based both on the textual content and the document structure. Finally, we introduce *semantic overlay networks*, which support similarity search on information represented in hierarchically organized and multi-dimensional semantic spaces. Topics that go beyond data representation and search are summarized at the end of the lecture.

KEYWORDS

peer-to-peer systems, structured overlay networks, semantic overlay networks, data management, distributed query processing, load balancing, schema mapping, document search

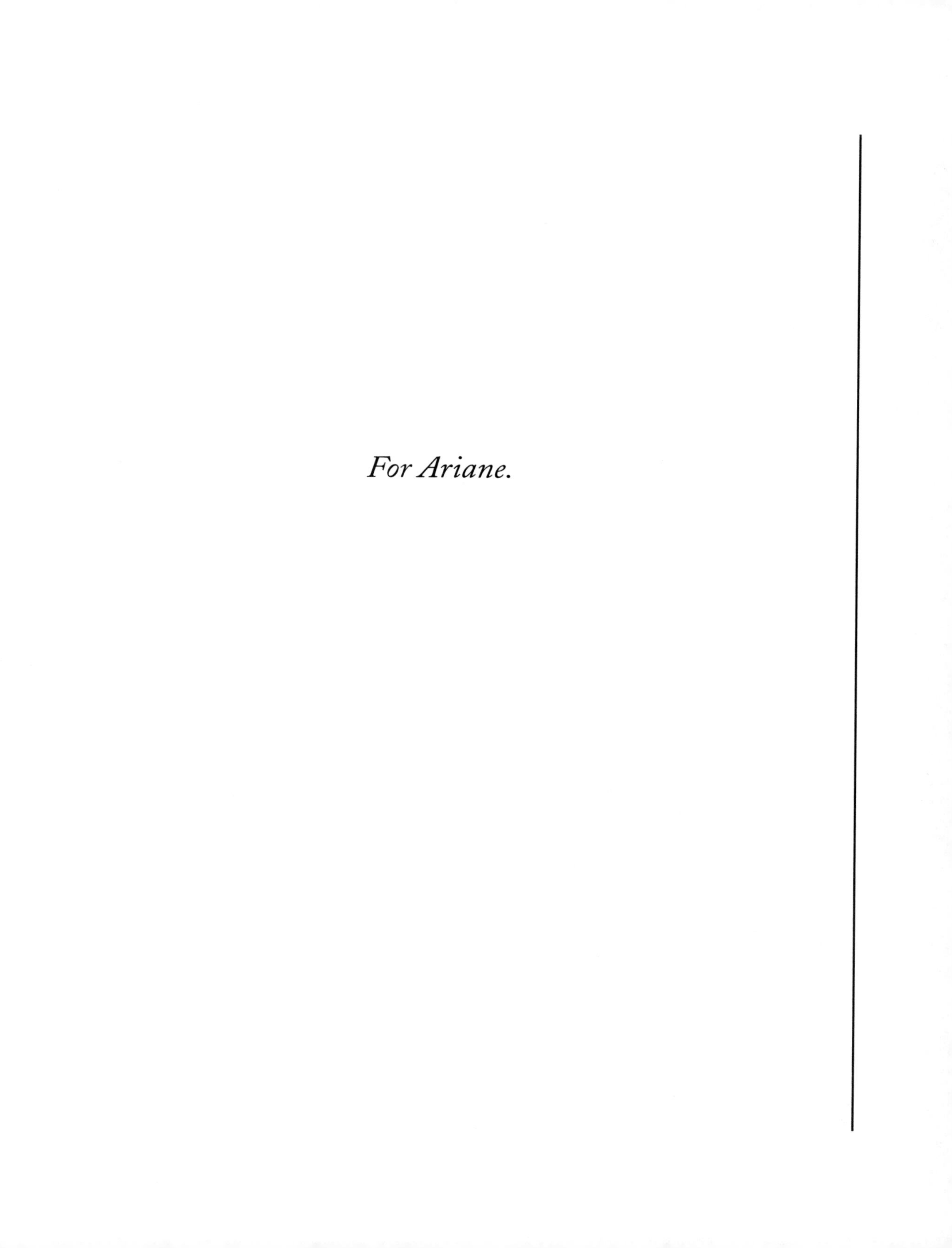

For Ariane.

Contents

Preface

Peer-to-peer data management connects the fields of peer-to-peer systems and data management. The nature of this connection is twofold. In peer-to-peer systems, search over large data sets is a key functionality; therefore, they have a natural link to the area of data management. The basic concepts for realizing scalable applications on the Internet that are at the core of peer-to-peer technology, on the other hand, inspired also novel large-scale data management architectures.

The focus of this lecture will be on the representation and search of structured data and documents in peer-to-peer data management systems. We adopt a wide perspective on the notions of data and search, including multiple facets of information management such as data integration, document management and text retrieval. This is motivated by the observation that these are key areas for which peer-to-peer architectures have been typically considered. In order to limit the scope of the lecture, we do not cover aspects related to state management and data dissemination, such as updates, transactions, data stream management, continuous querying and publish-subscribe. These areas are less mature today and will be shortly surveyed at the end of the lecture.

We will assume that the reader is familiar with the fundamentals of both the areas of peer-to-peer systems and data management. For peer-to-peer systems, there exist numerous surveys and for data management there exists a rich source of comprehensive textbooks. We will, in this book, recall from both fields the more specialized key concepts that will be necessary for the understanding of the specific techniques we discuss. These will be introduced in separate information boxes.

The lecture will cover four different types of peer-to-peer data management systems that are characterized by the type of data they manage and the search capabilities they support. The first type are *structured peer-to-peer data management systems* which support structured query capabilities for standard data models, in particular, relational data and Web data represented in RDF. The second type are *peer-to-peer data integration systems*. These support querying of databases that are using heterogeneous database schemas without requiring a common global schema. The third type are *peer-to-peer document retrieval systems* that enable document search based both on the textual content and the document structure, e.g., represented as XML. Finally, we will introduce *semantic overlay networks,* which in a sense generalize the previous types of systems. They manage data from both hierarchically organized and multi-dimensional semantic spaces that are equipped with a similarity measure to model semantic proximity.

Karl Aberer
May 2011

CHAPTER 1
Introduction

Peer-to-peer systems have generated a wide interest over the last decade, both in research and practice. Sharing of media content, in particular of music and video, through peer-to-peer systems has become a highly popular and the most bandwidth consuming application on the Internet. It has also induced huge debates on the legality of this activity and the economic model employed by the media industry. From a data management perspective, peer-to-peer content sharing systems pose interesting challenges, as they can be understood as very large scale and widely distributed databases. For example, the problem of searching for content based on a variety of criteria within a highly decentralized system poses interesting challenges that had not been addressed at that time. Initial practical solutions were based on simplistic albeit robust approaches, such as the famous Gnutella protocol. In parallel intensive research efforts in distributed systems, networking and data management started with the goal of designing more efficient solutions to accessing and managing data in a peer-to-peer architecture. This resulted in a rich set of novel algorithms and system architectures. Many of these address classical data management problems under a completely novel perspective. Scalability, minimizing communication cost and decentralized control are among the most prominent aspects that motivated this research. Over time, the focus expanded from the initial problem of content search, to many data management problems that can be studied in a peer-to-peer architecture. In this lecture, we will provide an overview of the most important principles, solutions and applications based on the peer-to-peer paradigm for data management problems.

In this chapter, we will first survey the basic notions of peer-to-peer systems on which we will rely in this lecture. Then we will introduce a taxonomy of peer-to-peer data management architectures to set the framework for the subsequent chapters in this lecture.

1.1 PEER-TO-PEER SYSTEMS

The notion of peer-to-peer systems is used in a wide variety of contexts and implies different technical concepts and approaches. In the following, we introduce a taxonomy of the key concepts that are relevant for the purpose of this book. We will also review some key technical approaches to peer-to-peer system architectures that peer-to-peer data management systems rely on. For more detailed technical discussions of peer-to-peer technologies that have emerged over the past 10 years, we refer the reader to one of the numerous surveys and books [Aberer and Hauswirth, 2002, Aberer et al., 2005, Androutsellis-Theotokis and Spinellis, 2004, Korzun and Gurtov, 2010, Lua et al., 2005, Milojicic et al., 2002, Oram, 2001, Risson and Moors, 2006, Vu et al., 2010].

1.1.1 CONCEPTS

In its most general sense, a *peer-to-peer system* is a computational system consisting of a set of *peers* P that jointly perform a common task and are connected to *neighbors* through *links* in a *logical network*. This logical network is called the *peer-to-peer overlay network*, or sometimes short peer-to-peer network, overlay network or simply network, when the context is clear. The peers communicate to their neighbors through a *communication network*, which is called the *underlay network*. In the literature, peers are also frequently called nodes or agents, but for the sake of consistency, in this book, we will always use the term peer. Links are often called references, connections or vertices, but we will consistently use the term link. The communication network is typically the Internet, but other means of communication like wireless communication networks may be considered. Even communication among peers that are hosted on a common server might be considered as a realization of the underlay network.

In the ideal case, different peers are assumed to have *equivalent roles*, state is distributed among the peers and control is *decentralized*. In other words, there exists no entity that has access to the global state of the system or exerts global control. As a result, decisions are always based on *local information* of peers, and the global system behavior is an *emergent property* of the system. Peer-to-peer systems are also often considered as *self-organizing* systems. The nature and formalization of the concepts of *emergence* and self-organization is an intensely debated topic into which we will not enter here [Serugendo et al., 2004].

The first essential function of a peer-to-peer system is *overlay network maintenance*. Every peer $p \in P$ maintains a set of neighbors $N(p) \subseteq P$ of which it knows their physical address in the underlay network. The data structure used to store the neighbors, their physical addresses and potentially other information about the neighbors is called the peer's *routing table*. Different overlay network architectures differ in the structure and contents of their routing tables. A characteristic property of peer-to-peer networks is the small size of the set of neighbors as compared to the *network size* $n = |P|$, typically constant or logarithmic in n. This property implies that the state of a peer-to-peer network is necessarily distributed since no peer has a global view of the peer-to-peer network.

Peers can autonomously decide to join or leave a peer-to-peer network. The process of joining and leaving is called *churn*. For joining a network a peer $p \in P$ needs to contact a peer $p_c \in P$ that is currently part of the network. Therefore, the join operation has the signature $join(p, p_c)$. Peers can always autonomously leave the network; therefore, the leave operation is of the form $leave(p)$. Overlay network maintenance protocols ensure that under churn the overlay network structure is maintained, i.e., that peers have routing tables that satisfy the constraints imposed by the overlay network architecture. The overlay network maintenance protocols are distributed, thus resulting in decentralized control of the peer-to-peer overlay network.

Peer-to-peer systems have been designed to perform a variety of tasks. However, most frequently, they are used to share and provide access to a set of *resources* R. These resources can be information resources such as documents, media contents, data, knowledge or services, or physical resources. Examples of physical resources are storage space, computational capacity or network bandwidth. We will call such systems *peer-to-peer resource sharing systems*.

Resources in a peer-to-peer system are either identified by a unique *resource identifier* from a *resource identifier space* $I(R)$ or by *resource properties* represented in application-specific data structures. A peer-to-peer resource sharing system supports basic functions for *resource management*. Basic examples of such functions are the following:

- $insert(p, r)$: add a resource $r \in R$ at peer $p \in P$.
- $lookup(p, id(r)) : p_r$: request from a peer $p \in P$ the physical address of a peer $p_r \in P$ that holds a resource with identifier $id(r) \in I(R)$.
- $query(p, q) : P_q$: issue a query q at peer $p \in P$ returning the physical addresses of a set of peers $P_q \subseteq P$ holding resources that satisfy the conditions of the query.

The access to the resources at a peer is realized through application-specific protocols. In order to implement these functions, the peer-to-peer networks use *routing protocols* for locating peers that hold specific resources. Routing protocols are distributed algorithms that use the overlay network structure to exchange messages among peers. The overlay network structure together with the routing protocols are the essential constituents of a peer-to-peer system architecture.

A special type of resource are the peers themselves. They are identified by *peer identifiers* from a peer identifier space $I(P)$. The same set of operations as provided for general resources can be applied to peers. For example, the lookup operation would have the form $lookup(p, id(p_s)) : p_s$ and return the physical address of a peer given its identifier. Figure 1.1 illustrates the concepts introduced.

1.1.2 ARCHITECTURES

Peer-to-peer overlay networks are the essential constituent of a peer-to-peer system. Any function of the peer-to-peer system implemented as distributed algorithm makes use of the overlay network, and any state information that is present in a peer-to-peer system is accessed through the overlay network. There exist three main types of overlay network architectures:

- Unstructured overlay networks
- Structured overlay networks
- Hierarchical overlay networks

We introduce in the following the main principles for each of these architectures and provide a canonical example for each type.

Unstructured overlay networks. In an unstructured overlay network, the choice of neighbors of a peer is not constrained by the state of the peers. In other words, peers can freely connect to any other peer in the network, by obtaining its physical network address. Similarly, also the choice of

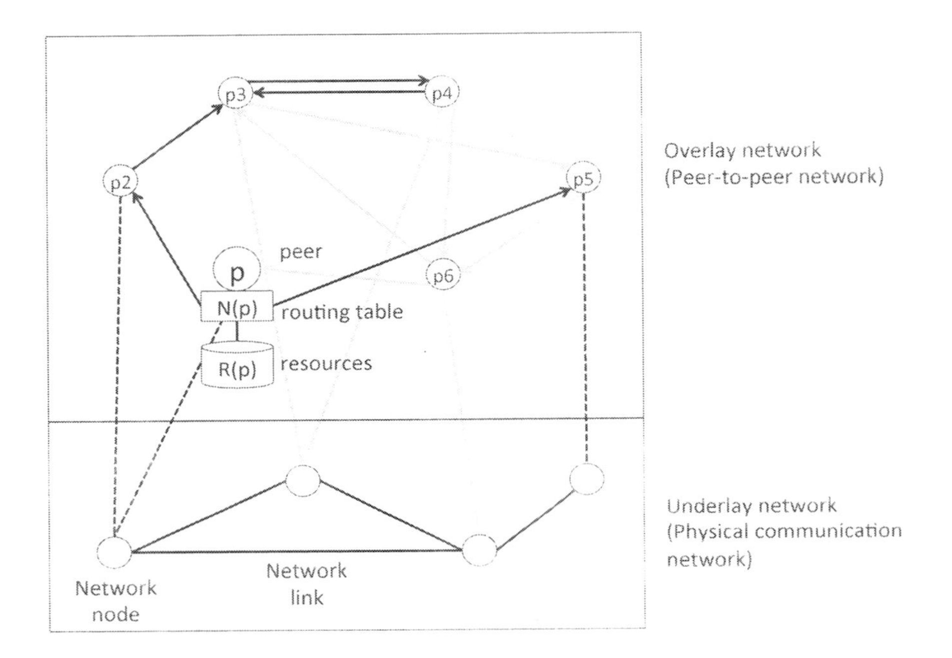

Figure 1.1: This figure summarizes some essential aspects of a peer-to-peer network. The peers in the peer-to-peer network are connected by logical links. Each peer, such as peer p, has a routing table with information on the peer's neighbors and resources that it manages. A logical link may correspond to a path in the underlying communication network. For example, the link between peer p and peer p_5 corresponds to two physical hops in the communication network. Conversely, peers that are separated through multiple links in the overlay network, may be physically close, e.g., peers p and p_3. Different peers may be hosted in the same physical node, for example, peers p and p_2. The overlay network may contain cycles, e.g., between peers p_3 and p_4.

which resources a peer can manage is not constrained by the state of the peer. In a pure unstructured overlay network, a peer does not maintain any information about the state of other peers in its routing tables. However, this assumption is frequently relaxed for optimizing the performance of unstructured overlay networks. As a result of these architectural principles, when searching a resource in the overlay network, typically, a large fraction or the whole network needs to be accessed.

Gnutella. Gnutella was the first unstructured peer-to-peer system used in practice and the starting point for a family of peer-to-peer protocols based on *gossiping* [Ripeanu, 2001]. The network structure consists of a directed graph with constant out-degree. For searching, a *breadth-first search strategy (BFS)* is adopted: a query message is forwarded to all neighbors, who, in turn, forward it to their neighbors till the message has been forwarded a *time-to life (TTL)* number of steps. Peers remember the recently sent messages to avoid cyclic forwarding. If a result is found, a query reply message is sent back to the query originator along the same path the query message has been forwarded. Network maintenance relies on a similar approach. A peer that is joining the network or in need of new neighbors announces its presence by broadcasting a *ping message* to the network, using the BFS strategy. Any peer receiving a ping message may decide to respond with a *pong message* that is routed back along the path the ping message arrived.
Several improvements have been proposed to decrease the high message traffic incurred by the BFS protocol [Lv et al., 2002, Tsoumakos and Roussopoulos, 2006].

- *Selective forwarding.* A search message is only forwarded to a fraction of the known neighbors. If this fraction is chosen carefully, still the whole network can be reached. This approach is called *percolation search* [Sarshar et al., 2004].
- *Iterative deepening.* The search starts with a low TTL. Only if no results are found, a new search with higher TTL is initiated. This approach reduces message traffic in particular when searching frequently occurring resources.
- *Random walkers.* Instead of relying on breadth-first search, this strategy uses *depth-first search*. A random walker selects among the neighbors uniformly at random one peer to forward the query message. In this approach, the TTL is chosen significantly higher. One or more parallel random walkers can be used.

The main characteristics of unstructured overlay networks can be summarized as follows. They typically incur high message traffic for performing searches, while the maintenance of the overlay network and the update of resources is less costly. They are robust against failures since for their operation no specific constraints, apart from avoiding disconnection, need to be satisfied. Their search mechanism does not limit the expressivity of queries used, as queries are evaluated locally at each peer.

Structured overlay networks. In this architecture, the structure of the overlay network as well as the assignment of responsibility of peers to manage specific resources follows a global organization principle. The organization of a structured overlay network is based on the association of the peers and the resources with some common, application-specific identifier space I, in other terms, $I = I(P) = I(R)$. Domain-specific identifiers of resources and peers are mapped to the identifier space using hash functions $h_R : R \rightarrow I$ and $h_P : P \rightarrow I$. For resources the domain-specific identifiers consist typically of some metadata, whereas for peers they are peers' IP addresses.

The identifier space I is equipped with a topology; typically, it is a one or multi-dimensional metric space. Both, the selection of neighbors of a peer and the responsibility to manage certain resources, depend on the peer identifiers. The set of resource identifiers a peer is responsible for is defined as a local neighborhood of the peer identifier. The routing tables consist usually of *short-range links* to direct neighbors in the peer identifier space, which ensure basic connectivity of the network, and *long-range links* to remote peers, which accelerate resource lookup by resource identifier, when those resources are found in a distant region of the identifier space. The significant difference to unstructured overlay networks is that lookups are performed by *greedy routing*. A lookup message, searching for a resource identifier, is always forwarded to the neighbor that reduces the distance to the target the most. Thus, searches do not flood the whole network. On the other hand, maintenance of a structured overlay network under churn may become costly since the structural constraints need to be maintained.

Numerous structured overlay network approaches have been proposed and studied over the past years. Here we distinguish these approaches primarily by the type of identifier space they employ and mention some of the well-known, canonical approaches.

- *One-dimensional ring.* The identifiers are binary strings of fixed length l. They are arranged in a ring topology with distance metrics $d(id_1, id_2) = |id_2 - id_1| \mod 2^l$, $id_1, id_2 \in I$. The most prominent representative of this approach is *Chord* [Stoica et al., 2001].

- *Binary prefix trees.* The identifiers are taken from the set of bitstrings $\{0, 1\}*$. The distance metrics can be given by the XOR-metrics, i.e., $d(id_1, id_2) = id_1 \oplus id_2$, $id_1, id_2 \in I$, where $(id_1 \oplus id_2)_i = 1$ if $(id_1)_i \neq (id_2)_i$ and 0, otherwise. The most prominent representatives are P-Grid [Aberer, 2001] and Kademlia [Maymounkov and Mazières, 2002].

- *m-dimensional torus.* The identifiers are taken from an m-dimensional torus, typically $[0, 1]^m$. The metrics is the Euclidean distance. The most prominent representative of this approach is CAN [Ratnasamy et al., 2001].

For an extensive and detailed review of structured overlay network approaches, we refer in particular to Korzun and Gurtov [2010].

Chord. Chord has been among the first proposals of a *distributed hash table* [Stoica et al., 2001]. It is based on *consistent hashing*, which applies a uniform hash function, such as SHA-1, to map resources and peers into a common identifier space. The identifier space in Chord consists of fixed-length binary strings of length l on a one-dimensional ring. The distance on the ring is computed modulo 2^l. By virtue of its hashed identifier, a peer p becomes responsible to manage the resources that are mapped into the interval $[p, successor(p))$, where $successor(p)$ is the peer with the next higher identifier on the ring. Peers are connected to their immediate successor which provides basic routing capability in the overlay network. In addition, peers maintain long-range links in exponentially increasing distances. More precisely, Chord requires a peer p to maintain links to the peers $successor(p + 2^i - 1), i = 1, \ldots, l$. As a result greedy search succeeds in $O(\log n)$ steps. For maintenance, peers that join have to obtain their long-range links and at the same time other peers may have to update their routing tables as a result of the join operation. The cost for a join operation is $O(\log^2 n)$. Chord has been the basis for numerous structured overlay network designs based on variations of its principles [Korzun and Gurtov, 2010].

P-Grid. P-Grid was an early proposal of a structured overlay network for supporting data-oriented applications [Aberer, 2001, Aberer et al., 2002b]. P-Grid uses an order-preserving hash function to map data into variable-length binary strings. Peers are logically organized in a binary tree. A peer with identifier p corresponds to the leaf node of the tree, which is reached by the binary path p. In its routing table, it maintains links to peers that correspond at each level of the tree to the opposite half of the subtree the peer belongs to. Formally, a peer with identifier $p = b_1 \ldots b_l$ maintains for each prefix $b_1 \ldots b_{l'}, l' < l$ a link to a peer with identifier prefix $b_1 \ldots (1 - b_{l'})$. Searching for a key, it then proceeds by identifying in the routing table the peer with the longest shared prefix and forwarding the search request to it. For balanced trees, this implies by construction that the search cost is bounded by $\log_2 n$, and it has been shown that, also, for unbalanced trees, the search cost is $O(\log n)$. Similarly as for Chord, a peer joining can build up its routing table in $O(\log^2 n)$. Other approaches based on prefix routing have been Kademlia [Maymounkov and Mazières, 2002], which introduced the concept of XOR metric, and Pastry [Rowstron and Druschel, 2001] and Tapestry [Zhao et al., 2004], which use higher-degree trees as underlying abstraction.

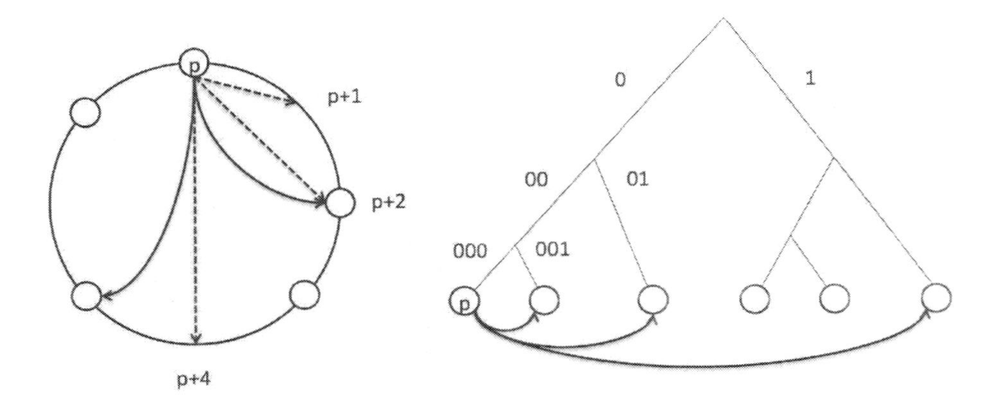

Figure 1.2: Illustration of the overlay network structure of Chord and P-Grid.

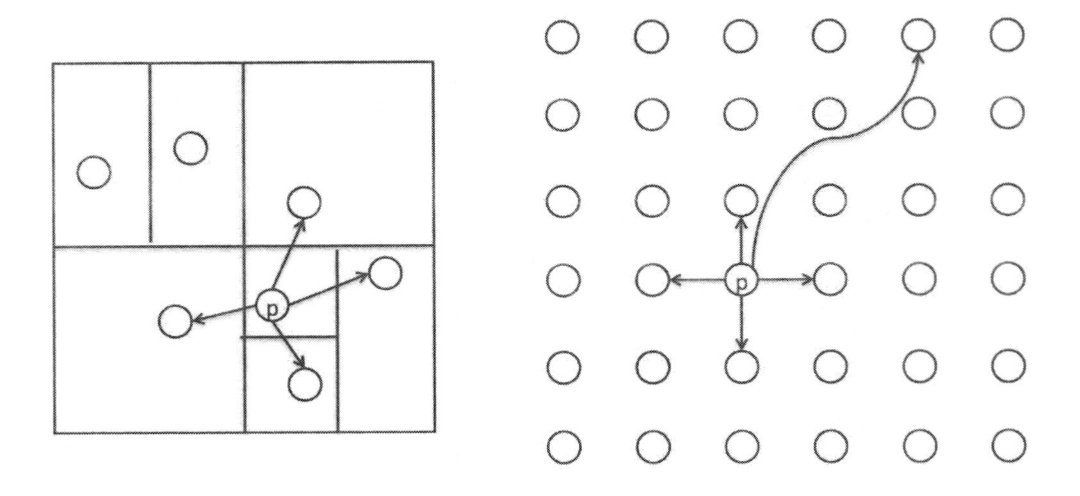

Figure 1.3: Illustration of the overlay network structure of CAN and small world networks.

CAN. Content-adressable networks (CAN) was the first proposal using a multi-dimensional identifier space for structured overlay networks [Ratnasamy et al., 2001]. Identifiers are taken from m-dimensional torus $[0, 1]^m$. The identifiers are generated by applying a uniform hash function m times to the resources and peers. The m-dimensional torus is recursively split into *zones* till a peer is the sole occupant of a zone. This defines the resources it manages and the neighboring peers. The routing table maintains $2m$ links to the peers in directly neighboring zones in the m-dimensional space. Query messages are forwarded to the neighbor with coordinates that are closest to the target in Euclidean metric. Thus, the routing cost is $O(mn^{\frac{1}{m}})$. A joining peer has to first locate the peer responsible for its identifier. It splits the zone of the peer currently responsible for its identifier into two and establishes the routing table. Thus, the cost of a peer join is $O(mn^{\frac{1}{m}})$. The approach of CAN can be extended by including long range links in exponentially increasing distances, similarly as in Chord and P-Grid, in order to achieve logarithmic routing cost [Xu and Zhang, 2002].

For structured overlay networks substantial research has been conducted to study the network performance and stability under realistic network conditions. We do not have the space to discuss those here in more detail but mention only some of the typical questions that have been addressed:

- *Maintenance.* An important performance characteristics is the behavior of a structured overlay network under churn, i.e., nodes joining and leaving the network [Gummadi et al., 2003]. Since structured overlay networks impose strict structural constraints on the network, it may become difficult to maintain those in case the churn exceeds a certain rate. The maintenance cost might become elevated, or it might even be impossible to maintain a consistent network structure.

- *Load balancing.* Even when using uniform hash functions to distribute resources among peers, it occurs that the resulting load at the peers is non-uniformly distributed [Rao et al., 2003]. Various techniques have been proposed to remedy this situation, including virtual peers, the power of two choices [Byers et al., 2003] and the use of multiple hash functions.

- *Replication.* Since peers are often unstable, resources might become inaccessible if they are not replicated within the peer-to-peer system. Typical replication schemes store a resource not only at the peer responsible for the resource but also at some neighboring peers in order to increase availability [Datta et al., 2003, Lv et al., 2002]. Also *erasure coding* has been considered for that purpose [Datta and Aberer, 2006, Weatherspoon and Kubiatowicz, 2002]. Erasure codes encode data by adding redundant data such that the original data can be recovered from any subset of the encoded data of which the size is above a threshold. Erasure codes have the advantage that they can achieve the same failure resilience as replication with less redundancy.

Similarly, also the network links in the overlay network can be replicated in order to increase the failure resilience of the peer-to-peer overlay network itself.

- *Proximity-based routing.* Logical links in the overlay network might connect peers that are far apart in the underlay network, i.e., that are connected by long routes in the physical network. With proximity-based routing, the structure of the overlay network is aligned with the underlay network, such that neighboring links in the overlay network have also short connections in the physical network [Rowstron and Druschel, 2001].

An interesting observation is that most of the structured overlay networks follow a common underlying principle. They are realizations of *small-world graphs* that efficiently support greedy routing [Kleinberg, 2000]. Kleinberg showed that for a network of nodes that are connected in a regular grid by judiciously introducing long-range links the routing cost can be minimized for the greedy routing protocol. The standard structured overlay networks satisfy this necessary condition. Indeed, there exist a number of proposals for structured overlay network construction that directly apply the principles of small world graphs, either by explicit construction [Girdzijauskas et al., 2010, Manku et al., 2003] or as byproduct of query routing protocols [Clarke et al., 2001, Galuba and Aberer, 2007].

Small world graphs. In this model, peers are assumed to be connected in a m-dimensional regular grid which constitutes a network of short range links. Manhattan distance is used as distance metrics d on the grid. Search is performed by greedy routing. In order to accelerate greedy routing, long range links are added. In the case where per peer a constant number of long range links is added, it can be shown that for a grid with dimensionality m a long range link $l(p_1, p_2)$ between peers p_1 and p_2 should be chosen with the following probability depending on their distance $d(p_1, p_2)$:

$$P[l(p_1, p_2)] \propto d(p_1, p_2)^{-m} .$$

Then the routing cost using greedy routing is $O(\log^2 n)$. By adding a logarithmic number of long range links, i.e., choosing the number of long range links proportional to $\log n$, the routing cost becomes $O(\log n)$ [Girdzijauskas et al., 2005]. This model explains the behavior of most of the standard structured overlay networks mentioned above, such as Chord and P-Grid, as well as CAN when extended with long range links.

Hierarchical overlay networks. In practice, peers have widely diverse performance characteristics. Many nodes are unstable and remain in the peer-to-peer system only for short periods. Some nodes exhibit more server-like characteristics, have more available resources and stay in the system for extended periods. Based on this observation, most of the widely used practical peer-to-peer systems, like Gnutella, Emule, and Skype adopted a hierarchical approach for the peer-to-peer system architecture [Rasti et al., 2006]. More powerful peers, so called *superpeers,* take over a larger part of the system workload and behave towards less powerful, regular peers as servers. In a basic superpeer architecture, a subset $S \subset P$ of all peers take the role of superpeers. The superpeers connect in an unstructured or structured overlay network in order to route global search requests. Regular peers connect to one or more superpeers and forward search requests and updates to their superpeer. The superpeers typically maintain a global index of all resources available at the regular peers associated with them. In addition, the regular peers associated with a superpeer may form among themselves a local overlay network to process search requests. Only if local requests cannot be successfully resolved, they are forwarded via the superpeer to the global superpeer overlay network. Usually, superpeer networks are organized in a two-level hierarchy, but extensions using multiple levels of superpeer networks have been studied.

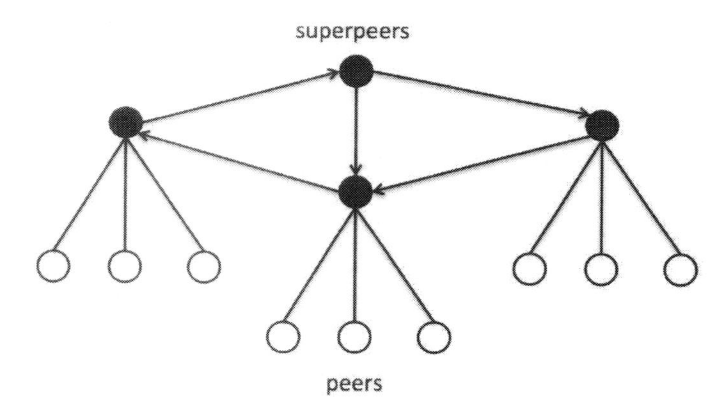

Figure 1.4: Illustration of a basic superpeer network. The superpeers are connected through a unstructured or structured overlay network.

Gnutella ultrapeers. In order to improve scalability of the Gnutella network, the ultrapeer concept has been introduced. An ultrapeer is a superpeer that is connected with other ultrapeers in a Gnutella unstructured overlay network. Different to the original Gnutella network, where each peer has an out-degree of typically 4, in an ultrapeer network, the out-degree is typically 30. Each ultrapeer maintains connections to up to 30 peers, and each peer attempts to connect to typically 3 ultrapeers. Thus, on average, 10% of the network consists of ultrapeers.

1.2 PEER-TO-PEER DATA MANAGEMENT

Peer-to-peer data management is a specific form of distributed data management, which considers large-scale distribution of data and a decentralized architectural approach. In a peer-to-peer data management system, at least one part of the system is implemented using a peer-to-peer architecture.

Historically peer-to-peer data management has several roots. The first root is found in the extension of basic peer-to-peer systems, as described in the previous section, with richer search capabilities. Since resource location is a central problem for peer-to-peer systems, the problem of search has been always at the heart of the development of those systems. Enhancing search capabilities from simple key-based lookup to structured and similarity-based queries was a natural development, which was undertaken typically in the distributed systems and networking communities. The second root can be attributed to the adoption of the peer-to-peer architectural concept for distributed, federated and Web databases. Realizing that this architecture bears the potential of building more robust and scalable large-scale systems, the database and Semantic Web community started to apply it to generalize data integration and Web data management architectures. As a result, also novel overlay network structures derived from conventional database index structures were developed. Finally, a third root can be identified in the field of distributed information retrieval. Given that Web scale retrieval requires highly scalable solutions, it was a quite natural step to consider peer-to-peer architectural concepts for this problem.

Given the broad interest peer-to-peer data management has received in different communities and contexts, it is not surprising that the peer-to-peer approach has been applied to a wide range of data management problems. This includes standard, structured query processing, similarity search, data integration, continuous query processing, stream data processing, transaction processing, publish-subscribe and content distribution, just to mention the most important ones. In this book, we will cover a substantial fraction of these problems, but given the limited space, we have to omit some. We keep the focus on traditional ad-hoc data access, whereas different forms of active data and content distribution as well as support for transactions and workflows will not be covered.

Peer-to-peer systems have been developed to build large-scale distributed systems relying on decentralized control and self-organization. Applying this paradigm to data management implies a shift in some of the basic assumptions that are traditionally made in this field. This shift of assumptions also lays the ground to the development of novel techniques. We summarize and illustrate here some of the most important of these assumptions:

- *Large scale:* Distributed databases and information systems typically have been small to medium size involving up to a few hundred nodes. Peer-to-peer architectures consider systems with thousands to millions of nodes.

- *Decentralization:* Traditional distributed data management techniques typically include some form of centralized control, which may imply bottlenecks for a large scale system. Examples are root or coordinator nodes in distributed index structures, transaction coordinators in distributed transactions or global schemas in heterogeneous databases (see Chapter 4).

- *Communication cost:* In distributed data management, communication cost plays a less central role than in peer-to-peer systems, where it is considered the predominant cost. This aspect plays, for example, an important role in peer-to-peer retrieval where potentially long posting

to the data management layer. This property has been termed in the context of peer-to-peer data management as *network data independence* [Hellerstein, 2003].

We can distinguish two major classes of peer-to-peer data management systems, depending on how the data management layer is organized. In both cases, we assume that the data corresponds to some *data model*, which can be a structured or a multi-dimensional data model. We will call the data of such a data model the *data space* D.

In *homogeneous peer-to-peer data management systems*, the peers jointly manage one logical database $R_D \subseteq D$ that is physically distributed in the network. The different peers provide equivalent logical access to the physically distributed data R_D. The data objects in R_D are considered as resources managed by the peer-to-peer system. The data is mapped from the *data space* D to the peer identifier space I by means of a mapping $h_R : D \rightarrow I$. The peer-to-peer overlay network of the peer-to-peer system provides the distributed access paths. This situation is depicted in Figure 1.5 on the left. As described in the previous section, the architecture of the peer-to-peer system consists of an overlay network on top of a communication network and peers are mapped to the identifier space I by means of a hash function $h_P : P \rightarrow I$. This type of system is the analog of traditional *distributed database system*. Homogeneous peer-to-peer data management systems will be the topic of Chapters 2 and 4.

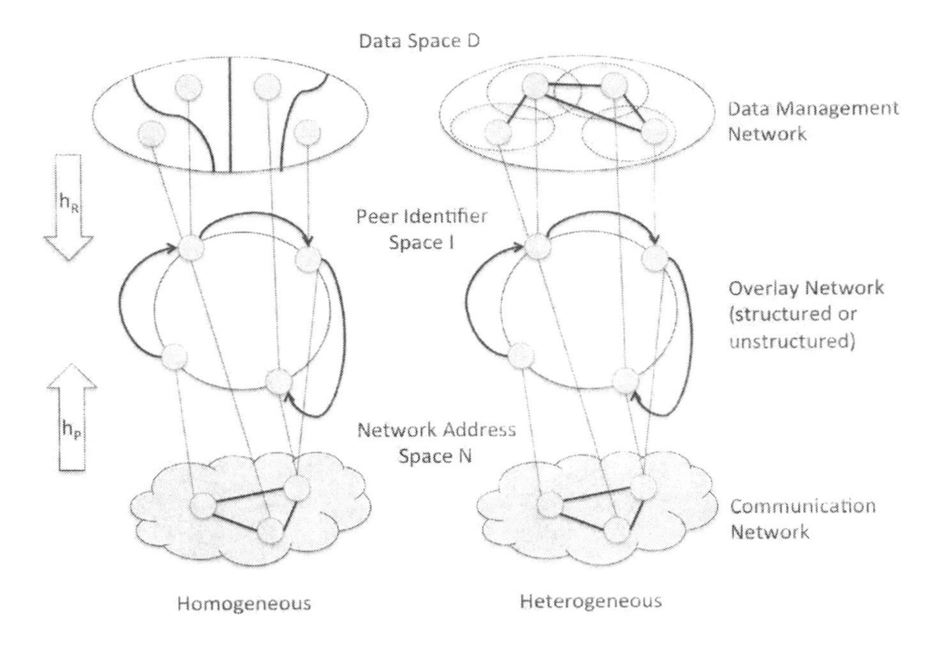

Figure 1.5: A generic architecture for peer-to-peer data management systems.

In a *heterogeneous peer-to-peer data management system,* the peers in the data management layer manage different logical databases, i.e., each peer p has a different *peer view* $R_D(p)$ on the data space. As a result, the same operation executed at different peers may produce different results, in contrast to the case of homogeneous peer-to-peer management systems. We depict this architecture in Figure 1.5 on the right-hand side. The peer view of each peer, corresponding to the subspace of the data space that it can access, is indicated by the ellipses around the peers. In addition, peers are connected to other peers in the peer-to-peer network at the data management layer. The peers form a logical data management network that is embedded in the data space.

A link in the logical network may indicate different kinds of communication capabilities among peers. In the simplest case, it corresponds to the ability to query and update data using a common protocol and data format. A link may carry information about the semantic similarity of the data managed by the peers. In more complex situations, a link may correspond to the capability of performing complex data processing, in particular mapping among different formats and types of data.

For the data management network, the data space D can play the analogous role as the identifier space I plays for a structured overlay network. The data space may be equipped with a distance or similarity metrics, similarly as for identifier spaces in structured overlay networks, and greedy routing is used as search strategy, but this needs not always to be the case.

In a heterogeneous peer-to-peer data management system, the relationship between the data management network and the overlay network, is analogous to the relationship among the overlay network and the physical network, for the peer-to-peer system. Different peers may be mapped to the same node in the overlay network and logical links among peers in the data management network may translate to multiple hops in the overlay network. Heterogeneous peer-to-peer data management systems are the analog of *federated databases* in traditional data management. Usually, only few heterogeneous peer-to-peer data management systems follow this architecture in its full generality. Frequently, different special cases of it are considered. These are shown in Figure 1.6.

Figure 1.6(a) shows the case where the structure of the data management network and of the overlay network coincide. This situation typically occurs when the identifier space of an overlay network coincides with the data space of an application. Figure 1.6(b) shows the situation where the peer-to-peer interactions take place only at a logical level, whereas physically the system is implemented in a single, centralized network node. Heterogeneous peer-to-peer data management systems will be the topic of Chapters 3 and 5.

1.2.2 DESIGN DIMENSIONS

For both the overlay network layer and the data management layer, a wide range of design choices is available. These design choices depend on the application requirements as well as on the performance objectives. We summarize in the following key design dimensions.

For the implementation of the overlay network structure in the peer-to-peer data management architecture, we can distinguish three approaches.

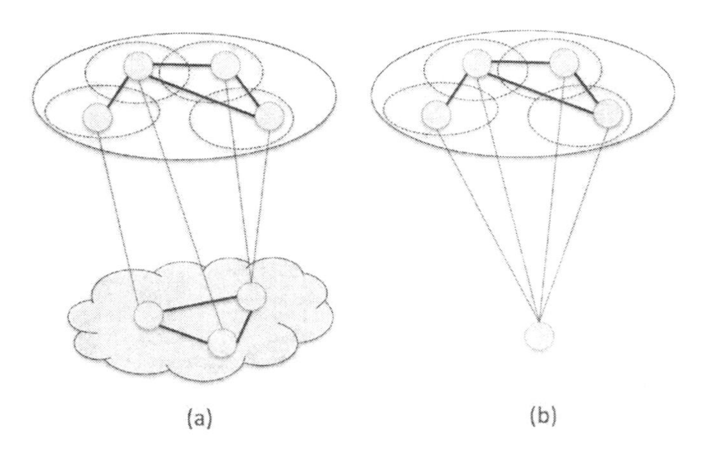

Figure 1.6: Special cases of the generic architecture for heterogeneous peer-to-peer data management systems.

1. *Use of existing overlay networks as black box.* In this approach, existing overlay networks are used in their original design. The advantage is that existing implementations and deployments can be used in their unmodified form. The disadvantage is that their performance characteristics or function might not be adapted to the needs of data management applications.

2. *Adaptation of overlay networks to data management requirements.* For the above reason in many approaches, existing overlay network designs are modified to better suit the needs of data management applications, like support for more complex query types with the resulting load balancing issues (see Chapter 2). A disadvantage of this approach can be that such modifications might make the design more complex or existing overlay network approaches do not permit to achieve specific design goals.

3. *Design of new overlay networks inspired by database indexing structures.* Thus, as a final alternative, novel overlay networks can be developed that are typically decentralized versions of classical database indexing structures, such as B+-Trees. Such overlay networks may indeed exhibit desirable properties from a data management perspective, but they may also sacrifice some principles of peer-to-peer systems, like requiring central control, and may suffer from high implementation complexity.

At the data management layer, the choice of the data model is a key distinguishing factor for different approaches to peer-to-peer data management. The data model is determined by the application needs and different data models may imply significantly different algorithmic problems that have to be tackled. We find for peer-to-peer data management the following typical data models.

- *Low-dimensional multi-attribute data:* This model is typically used for resource description and search. Data in this model can be understood as single relational table with multiple attributes. Typical operations are key-based lookups, value-based searches, range searches, conjunctive queries on multiple attributes and aggregation queries.

- *Structured databases with schema:* This is the standard data model used in traditional database systems, with the relational data model and XML as the main representatives. It can be understood as a generalization of the previous model. Data access is supported by full-fledged query languages, which implies in particular the capability to express join queries.

- *Semi-structured data:* This model is used in document management and Web data management, with RDF and XML as the main representatives. In this model, data is self-describing and tree- or graph-structured. A typical additional processing requirement as compared to the previous models results from the need to support path or graph pattern queries.

- *High-dimensional feature data:* This model is typically used for content retrieval. Data items are points in high dimensional metric spaces. Typical operations in this model are similarity searches, such as nearest neighbor or top-k queries.

Each of these models has been considered for homogeneous peer-to-peer data management systems. For heterogeneous peer-to-peer data management systems, we find two main variants.

- *Structured data with peer mappings:* In this approach, peers have different views on the data space by using structurally different representation for the same type of data. Links among peers correspond to the capability of mapping among different structural representations. This type of peer-to-peer data management system is called *peer-to-peer data integration systems.*

- *Multi-dimensional data with peer similarity:* In this approach, peers have different views on the data space by specializing on and covering certain subregions. These subregions correspond to interest or expertise profiles of peers. Links among peers correspond to forward queries to other peers that have similar profiles. This type of peer-to-peer data management system is called *semantic overlay network.*

1.2.3 OVERVIEW OF THE LECTURE

The forthcoming four chapters of the lecture will introduced four different types of peer-to-peer data management systems that are characterized by the type of data they manage and the search capabilities they support.

In Chapter 2, we discuss *structured peer-to-peer data management systems* which support structured query capabilities for standard data models. This type of peer-to-peer data management systems falls into the class of homogeneous peer-to-peer data management systems. The data these systems manage is either relational or RDF structured data. A key problem is the partitioning scheme that is used to distribute the data over the peers. Operations that are supported in this type of systems range from simple multi-attribute queries to complex join and aggregation queries.

The following Chapter 3 is dedicated to *peer-to-peer data integration systems*. These support querying of databases that are using heterogeneous database schemas without requiring a common global schema. They belong to the class of heterogeneous peer-to-peer data management systems. The data in these systems is typically relational, XML or RDF. The key problem for these system is the establishment and use of mappings to overcome heterogeneity among different schemas. These systems support usually full-fledged structured query languages.

Next, in Chapter 4, we focus on *peer-to-peer document retrieval systems* that enable document search based both on the textual content and the document structure, e.g., represented as XML. This type of system is another frequently encountered example of homogeneous peer-to-peer data management systems. A key problem for this type of system is to efficiently index document features, both text and structural features. Since these can be highly non-uniformly distributed, this poses significant challenges for load balancing and keeping communication cost acceptable when searching based on those features.

Finally, in Chapter 5, we introduce *semantic overlay networks,* which in a sense generalize the previous types to systems that manage data from both hierarchically organized and multi-dimensional semantic spaces that are equipped with a similarity measure to model semantic proximity. They are another example of heterogeneous peer-to-peer data management systems. A key question for this type of systems is to find mappings from the data space to the identifier space of peer-to-peer overlay networks. Since, in the data space, proximity is used to represent semantic similarity, these mappings should be proximity-preserving. This allows the system to efficiently handle similarity-based queries, as typically needed for content retrieval and information search.

We conclude the lecture in Chapter 6, by highlighting topics that we have not addressed in this lecture and that either belong to the realm of peer-to-peer data management or are closely related to it.

CHAPTER 2

Structured Peer-to-Peer Databases

Managing structured data in peer-to-peer systems aims at extending basic key lookups to more complex search types. The peer-to-peer system manages jointly one logical distributed database, and each peer supports the processing of complex queries. Such a system is an instance of a homogeneous peer-to-peer data management system. Complex query support in peer-to-peer systems is of interest both from a practical and theoretical perspective. Searching for resources by using complex conditions on metadata has clearly practical use in many peer-to-peer applications. Studying distributed data management architectures based on principles of decentralization and self-organization is also of a theoretical interest.

Depending on the type of overlay network used and on the type of queries, the technical challenges for implementing structured peer-to-peer databases are quite different. Exact key lookup queries can be efficiently answered using structured overlay networks. However, considering the simple extension to range queries already poses serious technical challenges since standard structured overlay networks do not maintain the order of hashed values when using uniform hash functions. For unstructured overlay networks, range queries pose no specific problem, as the query predicates are locally evaluated.

In the following, we will mostly focus on the case of using structured overlay networks to efficiently process complex queries. We start with the case of a single relational table that is horizontally partitioned among peers. We study the problem of efficiently processing range queries against single attributes and then generalize to processing multi attribute queries. Then we consider the extension of this approach to more complex relational queries, including join queries and aggregation queries. Finally, we will have a look at data models beyond relational, in particular, the RDF data model. Becoming the standard structured data model for Web data peer-to-peer solutions for RDF data management received particular attention.

2.1 RANGE PARTITIONING

Structured overlay networks were initially developed with the design goal of supporting exact key lookups of resources. Resources in R are mapped by means of a hash function $h_R : R \to I$ to an identifier space I. h_R is given as part of the overlay network design. Also, the peer identifiers are generated by hashing some property of the peer, typically its IP address, into the identifier space I. As a result of this mapping into a common identifier space, peers are assigned responsibility for managing resource identifiers in a subspace related to their peer identifier. For example, when I is a one-dimensional ring such as in Chord, the subspace managed by a peer $p \in I$ is $[p, successor(p))$, where $successor(p)$ is the peer with the next larger identifier on the ring. Structured overlay networks support typically the lookup of peers and resources by their hashed identifiers, which is the reason they are also called distributed hash tables. A distributed hash table manages thus key-value pairs $(h_R(r), r)$, where the key is the hashed identifier of a resource $r \in R$ and the value the resource itself.

Here we consider the case where the set of resources R is the set of tuples in a relational table $R(a_1, \ldots, a_k)$. A straightforward application of structured overlay networks is to make the relational table efficiently searchable by one attribute a_j. This is achieved by mapping the attribute to the identifier space I using a hash function h_R. As a result of this hashing, the tuples of the relation are distributed over the peers. This corresponds to a *hash partitioning* of the relational table [DeWitt and Gray, 1992]. This approach supports efficient key-based lookup of tuples distributed in the peer-to-peer network, i.e., efficiently evaluating the condition $a_j = v$ for some value v from the domain V of a_j.

In practice, other types of queries need to be supported, in particular when a_j is not used exclusively as identifier, but carries other meaning.

Example 2.1 A typical example using complex queries in the context of classical peer-to-peer application is the lookup of resources or services querying metadata on their characteristics related to performance and cost. Assume a peer-to-peer system for sharing files. The files are described by a relation containing some metadata on the content. The structure of the relational table could be

$$R(filename, year, downloadspeed, popularity) .$$

Then even a simple search naturally requires the evaluation of complex predicates, for example

```
filename contains "Beatles"
  AND year < 1968
  AND downloadspeed > 128kBit
  AND popularity > 0.8
```

Such searches would not be supported by a distributed hash table using uniform hash functions. The randomized nature of the hash function destroys the inherent order relationships among application keys, as illustrated in Figure 2.1(a). Thus, the distributed hash table does not support efficient answering of containment and range queries. On the other hand, the identifier space used in the design of structured overlay networks is usually a one- or multidimensional metric space. It lends itself naturally for representation of data from ordered domains, such as numerical values, as in the previous example.

Therefore, in order to enable range queries on an attribute a_i, a natural idea is to use a hash function h_R that maintains the order relationship among keys. We call such a hash function an *order-preserving hash function*. An order-preserving hash function h satisfies the following property: for keys $k_1 < k_2$ the hash values satisfy $h(k_1) < h(k_2)$, i.e., the hash function is monotonic.

Without loss of generality, we will make in the following the standard assumption that a peer with identifier $p \in I$ is responsible for all resources with hashed attribute values a_j in the interval $[p, successor(p))$. Thus, the identifier space I is partitioned among n peers $p_1, \ldots, p_n$ into ranges $I_1, \ldots, I_n$. This partitioning induces a *range partitioning* $V_1, \ldots, V_n$ of the attribute space of a_j, where $V_i = [h_R^{-1}(p_i), h_R^{-1}(successor(p_i))$. The problem when using order-preserving hash functions is that attribute values in general are non-uniformly distributed, and, consequently, the resulting hashed values are also non-uniformly distributed in I, as illustrated in Figure 2.1(b). One of the key motivations to use a uniform hash functions for structured overlay networks was indeed to load balance the system.

Reconciling range partitioning of attribute values and load balancing in a structured overlay network led to the development a number of non-trivial techniques, which we will present in the following.

2.1.1 STATIC RANGE PARTITIONING

For addressing the problem of load balancing when using an order-preserving hash function, the simplest case is when the probability density function $f(v) : V \rightarrow \mathbb{R}$ of attribute values is known in advance and static. Then it is possible to define the hash function such that the resulting distribution of resource identifiers in I is uniform. Without loss of generality, let us consider the case of an one-dimensional attribute value domain $V = [0, v_{max}]$ and an one-dimensional identifier space $I = [0, id_{max}]$. Knowing $f(v)$ such a hash function $h_R : V \rightarrow I$ can be (approximately) constructed as follows:

$$h_R(v) = \lfloor D(v) \cdot id_{max} \rfloor$$

where $D(v) = \int_0^v f(x)dx$ is the cumulative distribution of attribute values. Such a hash function is called a *uniform order-preserving hash function*. It results in a uniform distribution of hashed values in the identifier space I, as illustrated in Figure 2.1(c).

Example 2.2 This approach has been used to load balance attribute values using a uniform order-preserving hash function for Chord [Cai et al., 2004]. For a Chord network with identifiers of bit length l, the hash function is given as $h_R(v) = \lfloor D(v)2^{l-1} \rfloor$. The use of static range partitioning has also been proposed for structured overlay networks based on tree topologies [Aberer et al., 2002b, Li et al., 2009]. For estimating the probability density function $f(v)$, statistics on samples of existing databases are used.

A fundamental limitation of using uniform order-preserving hash functions is that the identifier space I is discrete and cannot adapt to any arbitrary application key distribution. Assume the shortest interval that an identifier space I permits to represent is $\delta \cdot id_{max}$, $\delta > 0$ and all application

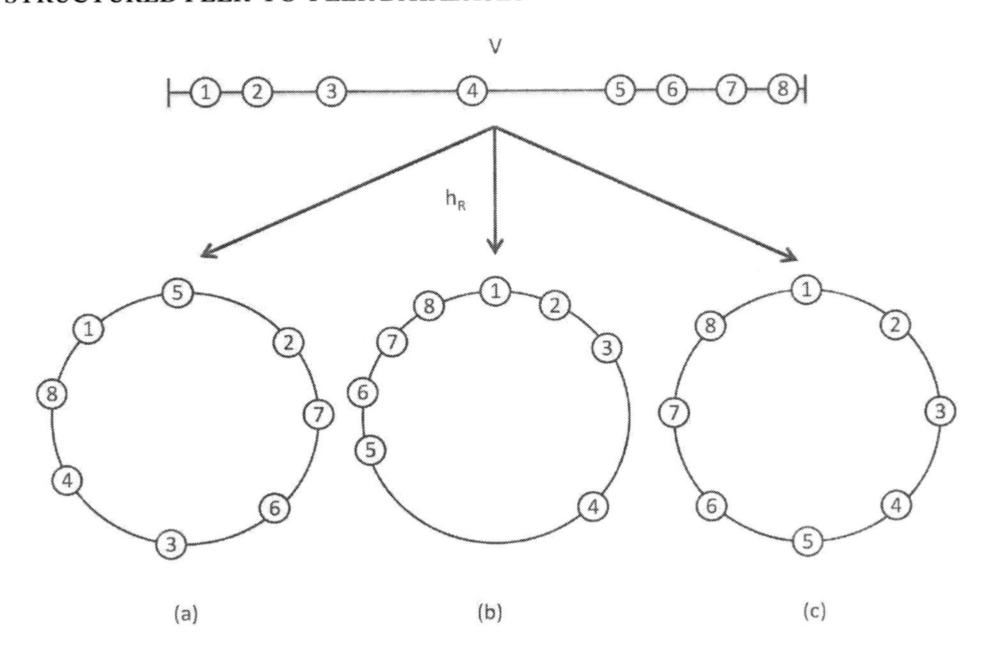

Figure 2.1: This figure illustrates the effect of (a) uniform hash functions, (b) order-preserving hash functions, and (c) uniform order-preserving hash functions on non-uniformly distributed attribute values from a domain V.

keys are concentrated in an interval of length $\delta \cdot v_{max}$. Then, necessarily, all keys will end up with at most two peers with consecutive identifiers in I. For example, for Chord $\delta = 2^{-l}$ and for a binary tree-based overlay $\delta = 2^{-l_{max}}$, where l_{max} is the length of the longest path in the tree.

Note that this limitation is not only of theoretical interest. For example, when using Web addresses of filenames as keys, it is possible that they all share a common prefix, which can be relatively long. When using 160 bit keys, as it is common in many DHTs, this could accommodate prefixes of a length of at most 20 bytes.

2.1.2 DYNAMIC RANGE PARTITIONING

In most practical applications, the distribution of attribute values is neither known in advance nor static. In this case, the peer-to-peer data management system has to dynamically learn and adapt the range partitioning to the attribute value distribution. Performing this task implies, in fact, solving three different problems:

1. Peers have to perform load balancing so that every peer manages approximately the same workload in terms of resources it is responsible for. In other words, the distribution of peer

identifiers has to align with the distribution of hashed attribute values. In absence of global coordination, peers have to perform this task in a decentralized way.

2. Peers have to establish a routing efficient overlay network in the presence of non-uniformly distributed peer identifiers as a result of the dynamic load balancing.

3. For performing both tasks, peers need statistics on the current load of the peers or the resource identifier distribution resulting from non-uniformly distributed attribute values. Learning precise global statistics might be expensive.

Note that these tasks are highly interdependent on each other. For example, the overlay network is often used both to collect statistics about the resources managed in the peer-to-peer network and to route to peers with which load balancing operations are performed. The load balancing operations again imply changes both in the statistics and in the overlay network structure. In the following, we will discuss in detail different approaches that can be taken to address these three tasks. We first look at the load balancing problem, assuming that already an overlay network structure, albeit imperfect, is in place.

2.1.3 LOAD BALANCING

We assume that the peers are connected in a one-dimensional ring topology, as in most structured overlay networks, and attribute values are mapped by means of an order preserving hash function to the identifier space I. In the following, we define the *workload* of a peer as the number of tuples that it manages. More precisely,

$$load(p_i) = |\{t \in R \mid h_R(t.a_j) \in I_i\}| \ ,$$

where $I_i \subseteq I$ is the range of identifiers managed by peer p_i, and a_j is the attribute in relation R that is used for indexing.

The basic idea for performing load balancing is that peers that are overloaded try to find other peers that are underloaded, and vice versa, and balance the load among each other. Also, peers that are joining the overlay network attempt to locate peers that have a high load and peers that leave the network try to offload their workload to underloaded peers. This gives rise to four possible load balancing operations:

- Join at high loaded peer: when a peer joins the overlay network, it can search for a peer with a high (or the highest) load and join at this peer.
- Leave and transfer to low loaded peer: when a peer leaves the overlay network, it can search for a peer with a low (or the lowest) load and transfer its load to this peer.
- Load balance with local neighbor: since peers know their local neighbors, they can mutually inspect their load and adjust it. This may imply that the peers change their identifier but not their relative position.

- Load balance with remote neighbor: if peers have information about the load of remote peers, they can use it to adjust their load with those. This implies that the peers change their identifier and their relative position.

Decentralized load balancing. Based on the four basic load balancing operations, a decentralized *threshold-based load balancing algorithm* can be devised [Ganesan et al., 2004a]. This approach assumes that complete information about peer load is available to all the peers. A peer attempts to shed its load whenever its load increases by a factor δ upon insertion of resources and attempts to gain load when it drops by the same factor. For that a geometric sequence of thresholds $\tau_m = \lfloor c \cdot \delta^m \rfloor, m \geq 1$ for some constant c is defined. When upon the insertion of a key, the load of a peer p_i increases beyond a threshold value, i.e., from τ_m to τ_{m+1}, the algorithm adjustLoad(p_i) given in Algorithm 1 is executed.

Algorithm 1: adjustLoad(p_i)

```
1  set m such that load(p_i) ∈ (τ_m, τ_{m+1}];
2  let p be the lighter loaded among p_{i-1} and p_{i+1};
3  if load(p) ≤ τ_{m-1};                          /* neighbor adjust */
4  then
5  |   move keys from p_i to p;
6  |   update identifiers of p_i and p;
7  |   adjustLoad(p);
8  |   adjustLoad(p_i);
9  else
10 |   Find the least-loaded peer p_k;
11 |   if load(p_k) ≤ τ_{m-2};                    /* reorder */
12 |   then
13 |   |   transfer all data from p_k to p_{k-1} and p_{k+1};
14 |   |   split data of p_i evenly between p_i and p_k;
15 |   |   update identifiers of p_i and p_k;
16 |   |   adjustLoad(p_{k-1});
17 |   |   adjustLoad(p_{k+1});
18 |   end
19 end
```

The peer p_i first attempts to load balance with its neighbors (neighbor adjust) if there is a neighbor that has less than $1/\delta$ of its load (lines 4-8). Otherwise, the peer chooses the globally least loaded peer p to attempt a load balancing operation with a remote peer (reorder) (lines 10-16). If p_k's load is small enough, less than $1/\delta^2$ of p_i's load, p_k distributes its load over its neighbors, changes its position in the overlay and splits the load with p_i. If p_k's load is too large, the system load is too high and no load balancing can be performed. Finally, the peers have to acquire their new identifiers corresponding to the resources identifiers they manage, and the procedure is recursively invoked for the peers that have changed their load.

Deletions of keys are handled symmetrically. When a peer's load drops to a threshold τ_m, it first attempts a neighbor adjust, provided the neighbors load is larger than τ_{m+1}. Otherwise, it attempts a reorder trying to split the highest loaded peer in the system if its load is more than τ_{m+2}.

The quality of a load balance algorithm can be measured by the *global load imbalance ratio*, defined as the asymptotic ratio between the maximal and minimal loads. A load balancing algorithm satisfies a global load imbalance ratio σ_c if for some constant $c \geq 0$ it guarantees

$$\sigma_c \geq \frac{\max_{p \in P} load(p) - c}{\min_{p \in P} load(p)} .$$

It has been shown that the threshold-based load balancing algorithm results in a global load imbalance ratio of at most $\sigma_c = \delta^3$ for any $\delta \geq (\sqrt{(5)} + 1)/2 \approx 1.62$, and that the amortized cost of the algorithm for insert and delete operations is constant.

The load balancing algorithm described above assumes that at any time the peers with the maximal and minimal load can be efficiently identified. One possibility to realize this is to maintain a separate order-preserving overlay network, where the value of the peer load is used as the peer identifier. Whenever the load of a peer changes due to insertion or deletion of resource keys as well as load balancing operations, this additional overlay network may potentially need to be updated, which might imply significant additional costs.

Load balancing with random sampling. Therefore, other methods to obtain statistical information about load distribution have been considered. One is to sample a number of peers in the network and then choose the one with the highest or lowest load, respectively, for performing the load balancing operation.

For the threshold-based load balancing algorithm, it has been shown that random sampling works as follows [Ganesan et al., 2004a]. If a particular peer wants to perform a reorder operation, it samples the network, and if it finds a peer that violates the threshold condition, it performs a load balance operation with that peer. Otherwise, it does nothing. In absence of data deletions, it can be shown that if the number of samples is $O(\log n)$, then the maximum load is at most within a constant factor of the average load with high probability.

Another approach is based on periodic sampling instead of threshold based triggering of load balancing operations [Karger and Ruhl, 2006]. Each peer performs occasionally Algorithm 2 at random using a load imbalance threshold $0 < \tau < 1/4$.

In this algorithm, when the peer contacts another random peer and the load imbalance ratio is beyond the threshold (line 1), it first checks if the next higher neighbor of the contacted peer has an even higher load than the peer itself. In that case, or if the both peers already are neighbors, they balance the load among each other (lines 8-9). Otherwise, the contacted peer p_k sheds its load to its neighbor, changes it position as new neighbor of p_i and balances its load with p_i (lines 11-13).

Algorithm 2: randomLoadbalance(p_i)

```
1  peer p_i contacts a randomly chosen peer p_k;
2  let load(p_i) ≥ load(p_k);                          /* swap indices if needed */
3  if τ · load(p_i) ≥ load(p_k) then
4      if |i − k| > 1 and load(p_{k+1}) > load(p_i) then
5          i := k + 1
6      end
7      if |i − k| = 1 then
8          share keys equally among p_i and p_k;        /* neighbor adjust */
9          update identifiers of p_i and p_k;
10     else
11         transfer all data from p_k to p_{k+1};       /* reorder */
12         split data of p_i evenly between p_i and p_k;
13         update identifiers of p_i and p_k;
14     end
15 end
```

Starting with an arbitrary load distribution, this algorithm has been shown to result with high probability in a state where peers have a load of at most $\frac{16}{\tau}L$, when peers contact $\Omega(n)$ other peers and where L is the average load of the system.

Load balancing with approximate statistics. It is also possible to find a middle ground between maintaining a complete global statistics and pure random sampling, by maintaining an approximate global statistics of the load distribution at the peers [Vu et al., 2009]. The method is generic and applicable to a variety of overlay networks. Each peer maintains a histogram of the average load of non-overlapping groups of peers. The non-overlapping groups partition the complete peer-to-peer network into groups of adjacent peers, and the peers maintain a link to one member of each group. In structured overlay networks it is usually possible to easily define such a partitioning using the routing tables, as illustrated in Figure 2.2 for the case of Chord. Whenever a peer receives a request for a search key, there is one unique entry in the routing table to which the search request is forwarded to. So each entry in the routing table corresponds naturally to one partition of the overlay network.

For load balancing using histogram information, the following property can be shown: if the maximum imbalance ratio between the load of a peer and the average load of a group of peers in its histogram is δ, the global load imbalance ratio is $\sigma_0 = \delta^2$.

A peer p that has a load $load(p)$ that is more than twice the average load of any of the groups in its histogram (i.e. $\delta = 2$) can always perform a load balancing operation in order to improve load balance in the system. Assume that there is group G in the histogram that has an average load smaller than $\frac{1}{2}load(p)$. One partitions G into pairs of peers adjacent in the key space, with one single peer

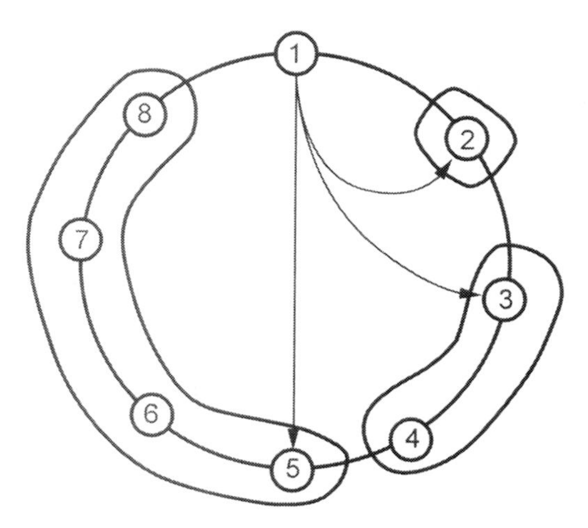

Figure 2.2: Example of partitioning a Chord network based on the routing table of peer 1. The three resulting partitions are {2}, {3, 4} and {5, 6, 7, 8}. For each, the average load is determined. Peer 1 connects to a group through the long-range link in its routing table.

remaining if the size of G is odd. Since the average load of the group is less than $\frac{1}{2}load(p)$, also, one pair or the single peer must have a load less than $\frac{1}{2}load(p)$. If it is a pair, then one peer can take over the whole load of the pair and the other peer half of the load of the overloaded peer. If it is the single peer p', there are two cases. If

$$load(predecessor(p')) + load(p') + load(successor(p')) < 2\,load(p)$$

then p' leaves its position and joins at p, otherwise it load balances with its direct neighbors and its load becomes larger than $\frac{1}{2}load(p)$. In all cases, the peer p is no more overloaded and the global load imbalance σ_0 in the system is $\delta^2 = 4$.

Conclusion. We can distinguish three different ways of how statistics is collected for load balancing in peer-to-peer networks.

1. Complete statistics is maintained and made available using a separate overlay network.

2. Peers sample the network to obtain an approximate statistics.

3. Each peer maintains locally an approximate global statistics.

Another feature that distinguishes different load balancing approaches is the events that trigger load balancing operations. These operations can be either triggered periodically by the peers, through join and leave events, or through the detection of local load imbalance resulting from updates.

2.1.4 OVERLAY NETWORK CONSTRUCTION

Dynamic range partitioning methods as described before enable peers to balance load among each other. But the resulting non-uniform peer identifier distribution violates the basic assumption made by most structured overlay network approaches that peer identifiers are uniformly distributed in the identifier space I. Let us illustrate the resulting problem by means of a simple example.

Example 2.3 Consider a Chord network with $n = 8$ peers and with $l = 5$ bit identifiers. Assume load balancing resulted in the peer identifier distribution as shown in Figure 2.3.

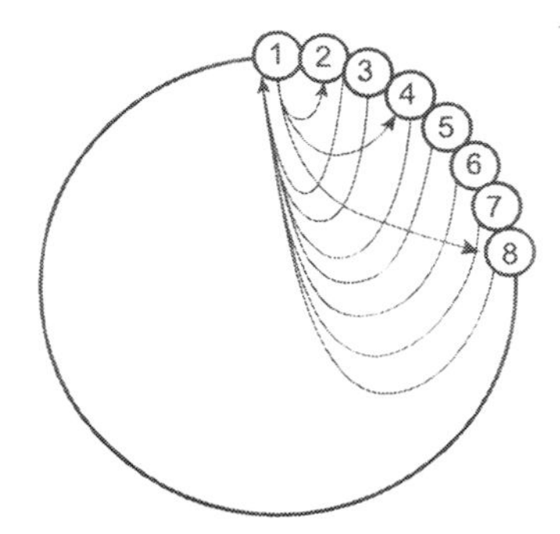

Figure 2.3: Outgoing and incoming neighborhood links for peer 1 with non-uniformly distributed peer identifiers.

The average routing cost in number of messages for searches starting at peer 1 is 1.5. This corresponds to the theoretical performance of Chord, which is given as $\frac{1}{2}log_2(n)$. However, for searches starting at peer 8 each search, except for the search for peer 8 itself, makes first a hop to peer 1 and continues then from peer 1. Thus, the average search cost is 2.25. This means that the network is less routing efficient than in the case of a uniform peer identifier distribution.

The construction of the routing topology for a structured overlay network with non-uniformly distributed peer identifiers has been approached in a number of different ways. In the following, we provide an overview of the main classes of approaches that have been developed for that purpose.

Sampling of peer identifier distribution. The first approach is based on the idea to learn the global distribution of peer identifiers. In the simplest case, peers could construct a global statistics of the peer identifier distribution and use it to derive a uniform order-preserving hash function as described in Section 2.1.1. However, building such a global statistics is not scalable for large networks as the cost would grow linearly in the size of the network.

Therefore, methods for obtaining an approximation of the peer identifier distribution have been considered [Bharambe et al., 2004]. The basic idea is that each peer samples the distribution locally, and the local estimates are exchanged in an epidemic style protocol.

A peer p first estimates locally the total peer count by contacting its ring neighbors $N_c(p)$ within a fixed distance c, e.g., $c = 3$. The local peer count estimate is determined as

$$n_{local}(p) = \frac{|N_c(p)|\, id_{max}}{\sum_{p' \in N_c(p)} |I_{p'}|} ,$$

where $|I_p|$ is the length of the interval in I that a peer p covers. We assume here that the identifier space is in the range $[0, id_{max}]$. Then peers randomly sample the network using k random walkers with a limited time-to-live exploiting the existing long range links of the overlay network in order to obtain a random subset of peers. Each of the k peers sampled in this way reports the e most recent peer count estimates of the form $(p, n_{local}(p))$ it has obtained from its own sampling process. Using these estimates, each peer constructs an approximate histogram of the global peer distribution. This histogram is used as an approximate probability density function of the peer identifier distribution in order to construct a constant number of long-range links in the following way: first a peer estimates from its histogram the total number n of peers in the system. Then, for each link, the peer draws a random number n_l using the harmonic probability density function

$$f_{harmonic}(x) = \frac{1}{n \cdot \log(\frac{x}{n})}, \quad x \in [1, n] .$$

Using the histogram, a peer estimates for which identifier id_l the peer at *hop distance* n_l is responsible for. The hop distance measures of how many successor hops are between two peers. Then it routes a search message for identifier id_l and establishes a long range link to this peer.

This sampling method to estimate peer identifier distribution works only for relatively well-behaved distributions, for example when frequency of keys depends monotonically on the key value [Girdzijauskas et al., 2010]. Figure 2.4 illustrates the difference between an simple parametric Zipf distribution and a typical distribution obtained from real-world data. Obtaining an accurate approximation of the latter by sampling is more difficult.

Therefore, a scalable sampling technique is needed that works well with any distribution [Girdzijauskas et al., 2010]. It uses the intuition that precise knowledge of the peer distribution in remote regions is less important for constructing long range links than for closer regions. In this approach, in a first step k random walks are performed in the network to obtain an estimate of the median of the distances to other peers in the whole peer population. Then a long range link is established to a peer chosen uniformly at random from the remote partition, i.e., from the set of peers that are farther away than the median distance. Within the closer partition again bounded random walks are performed, i.e., random walks that do not leave the closer partition. These determine the median value of distances of peers in the closer partition, and a long range link to the farther remaining partition can be established. This procedure is repeated till the remaining partition consists of the peer itself. During that procedure the TTL value of the random walks is set to the number of previously determined partitions. Therefore, in this approach the global number of peers needs not to be estimated.

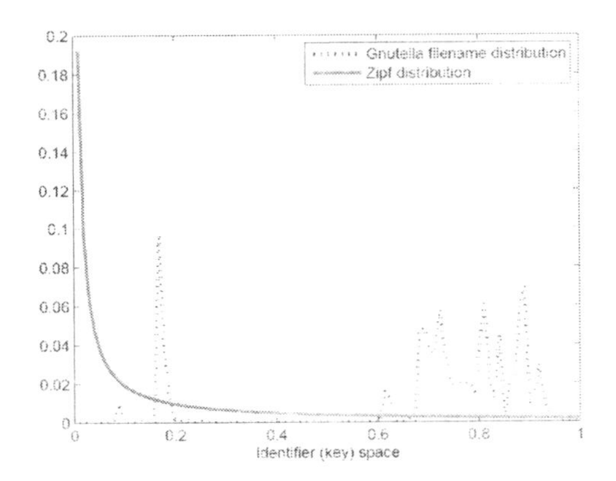

Figure 2.4: A regular distribution (Zipf) vs. an irregular distribution from real-world data, namely Gnutella filenames [Aberer et al., 2002b].

Hop count distance. In the case of static range partitioning we used a hash function that compensated for the non-uniform distribution of peer identifiers while maintaining their relative order (see Figure 2.1). A key observation is that the same effect can also be achieved when we consider the hop count distance between peers as distance measure for establishing long-range links. The sampling techniques described above indeed serve to estimate the number of peers that lie between two different peer identifiers. Instead of estimating this number by obtaining peer identifier distribution statistics, peers may also attempt to explicitly distinguish between an *identifier space*, where the distance between peers is determined as the difference of the value of their identifiers, and a *hop count space*, where the distance between peers is the number of successor peers that need to be traversed between two peers [Crainiceanu et al., 2007, Klemm et al., 2007, Schuett et al., 2008]. If the construction of the routing tables is purely based on the distance in the hop count space, the overlay network structure is insensitive to the distribution of peer identifiers in the identifier space. We describe here a variant that has been proposed as an extension of Chord [Crainiceanu et al., 2007, Schuett et al., 2008]. In a standard Chord network, peers construct their long range neighbors by contacting peers in exponentially increasing distances in the identifier space. In contrast, when using distance in the hop count space, long-range neighbors are obtained by the following incremental procedure. The first long-range neighbor of a peer is simply its successor peer. To determine its $i + 1$st neighbor, a peer contacts the ith neighbor in its routing table and requests its ith neighbor p. Then it links to p as its own $i + 1$st neighbor.

Discoupling attribute value and peer identifier space. SkipNet [Harvey et al., 2003] is a structured overlay network that distinguishes between two different spaces: an ordered attribute value space and a peer identifier space. Each peer is associated with an attribute value range and a randomly assigned binary string in the peer identifier space. The orders of attribute values and peer identifiers are not correlated. This is different to the previous approach using hop count distance, where the relative order of peers with respect to hop count distance and peer identifiers is the same.

The peers are bidirectionally connected in a ring in the order of the attribute value ranges they are associated with. This is the level 0 ring. Using the peer identifiers the routing tables are constructed in an hierarchical fashion. At level 1, two bidirectionally connected rings in the order of the attribute value ranges are constructed. Peers decide on whether they join the left or right ring depending on the first bit of their peer identifier. Analogously, the next higher levels are constructed till the rings consist of single peers only. The resulting overlay network structure is illustrated in Figure 2.5.

In their routing tables, peers maintain pointers to their left and right neighbors in the ring at each level. A search for an attribute value proceeds by identifying the highest level at which the ring neighbor does not point past the destination value. The search message is forwarded to this peer and the procedure is repeated till the peer responsible for the attribute value is found. This construction is insensitive to the attribute value distribution in the attribute value domain. It depends only on the relative ordering of the attribute values associated with peers. Therefore, it can deal with arbitrary load distributions resulting from load balancing among peers.

Routing by peer identifier is as well supported by this overlay network structure. It works analogously to prefix routing in tree-based overlay networks. A search starts by finding in the level 0 ring a peer that shares the first bit of the searched identifier. Then the search continues at level 1 resolving the second bit on the ring that contains all peers with a common first matching bit. This is repeated till the query is resolved.

Both search processes are illustrated in Figure 2.5. Searches for attribute values and peer identifiers can be performed in SkipNet with $O(\log n)$ messages. Interestingly, the same holds true also for the number of messages needed for a peer join [Harvey et al., 2003].

Distributed balanced trees. As an alternative to constructing routing-efficient overlay networks that consider non-uniformly distributed peer identifiers peer-to-peer implementations of database indexing structures based on balanced trees have also been investigated to support range queries in peer-to-peer data management.

Baton [Jagadish et al., 2005] maintains a balanced binary tree. In this approach, each peer is associated with a node in a binary tree, including the non-leaf nodes, and manages an attribute value range. The peers are ordered within the tree in order of the attribute values when performing a level order traversal. A peer maintains links to its immediate neighbors in the tree, the parent and child node, as well as to the adjacent nodes in the tree in inorder traversal order. In addition, at each level l of the tree the nodes are numbered from 0 to 2^l. The peers maintain long range links to nodes in the tree at exponentially increasing distances $1, 2, \ldots, 2^{l-1}$, both to the left and to the right. An example of a Baton overlay network structure is given in Figure 2.6. Using their routing tables containing long range links peers can detect cases, when the tree is imbalanced, i.e., when there exist two leaf nodes at depths that differ by more than 1. In such a case, the peers initiate load balancing operations that are comparable to rotations used for load balancing in AVL trees.

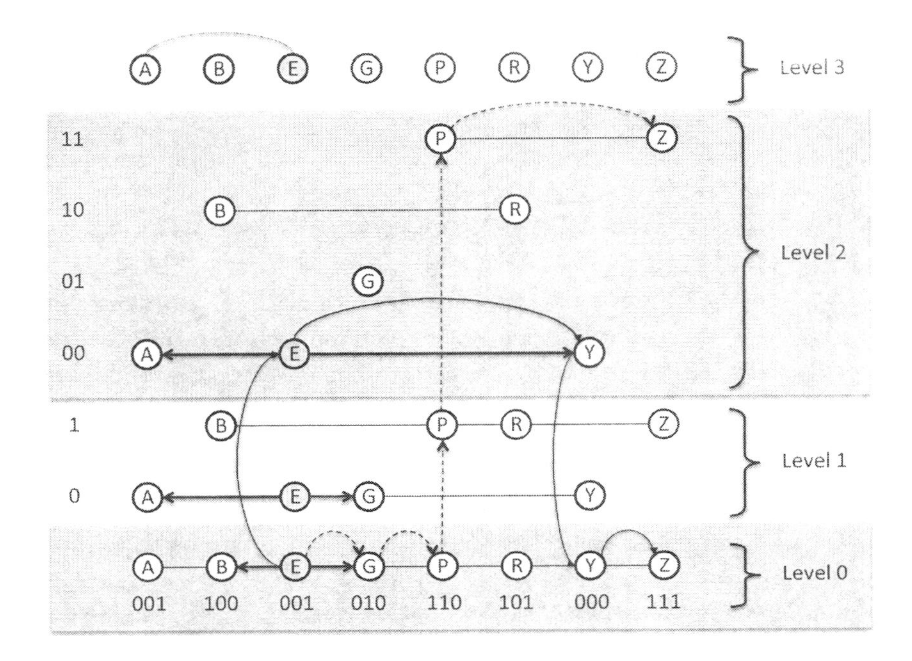

Figure 2.5: Example of a Skipnet. Note that the distribution of application specific identifiers does not influence the structure of the Skipnet, just their relative position. For peer E, the neighbor links at each level are indicated by bold arrows. The figure illustrates both an attribute value search and a peer identifier search starting from peer E. Note how for the attribute value search for value Y (plain arrows), the peer tries to identify the highest level at which the search can approach the target as fast as possible. In contrast, for a peer identifier search for identifier 111 (broken arrows), the peer tries to resolve the next bit of the searched identifier at the lowest corresponding level in the Skipnet.

P-Tree [Crainiceanu et al., 2004] distributes a B+-tree in a peer-to-peer network such that each peer maintains the left-most root-to-leaf path corresponding to its leaf position in the tree. The main challenge in this approach is to stabilize inconsistent views that peers might have on the global B+-Tree.

Conclusion. A wide variety of possibilities have been studied to construct routing efficient overlays for peers whose identifiers are non-uniformly distributed as a result of load balancing when using order-preserving hashing. These include:

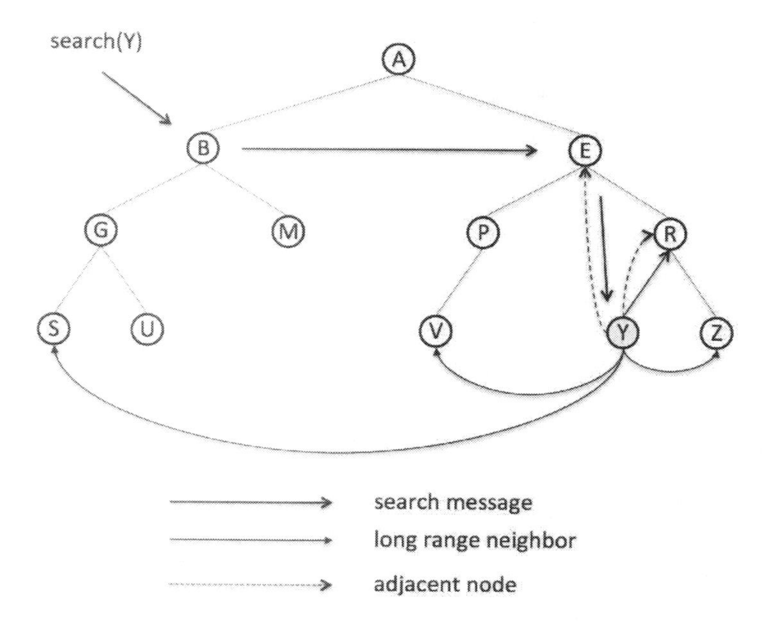

Figure 2.6: Example of a Baton network. For peer Y, the long range links and adjacent node links are indicated. A search for peer Y, that starts at peer B is first forwarded via long range link to peer E and then to peer Y, which is the node adjacent to E in inorder traversal.

1. Gathering statistical information about the peer identifier distribution and using it to choose correctly the long range links as described for the case of static range partitioning.
2. Determining the long-range links using hop count distance among peers instead of peer identifier space distance.
3. Decoupling the order of peer identifiers from the attribute value order.
4. Providing dedicated implementations of distributed, load balanced tree structures.

Each approach has different advantages and disadvantages in terms of search performance, maintenance and failure resilience. As of today, a comprehensive understanding of these tradeoffs is still missing and subject of future research.

2.1.5 EFFICIENT RANGE QUERY PROCESSING

When an order-preserving hash function is used, a simple strategy for processing range queries is the following: Peers search for the lower or upper bound of the query range I_Q of a range query Q and then traverse the adjacent peers till the whole query range is covered. The total message cost for an overlay network with typical search cost of $O(\log n)$ is then $O(\log n + |I_Q|)$. However, with this approach the query latency is also $O(\log n + |I_Q|)$. This can be problematic if $|I_Q|$ is large. Therefore, different methods for reducing the query latency in range query processing have been proposed. These methods exploit the structural properties of an overlay network to parallelize the retrieval of query results.

For tree-structured overlay networks, such as P-Grid and Kademlia, such a strategy is relatively straightforward [Datta et al., 2005]. In such an overlay network, peers are associated with a leaf node and maintain long range neighbor links at each level of the tree. The neighbor links connect to peers that are located in the opposite subtree at each level. When processing a range query a peer determines the highest level at which the corresponding subtrees intersect with the query range and sends in parallel messages to peers in those subtrees. The peers in the subtree recursively continue to process the query. The detailed processing for a range query $I_Q = [id_s, id_e]$ is given in Algorithm 3. $I_{p,l}$ denotes the range a peer p covers at level l of the tree. $neighbor(p, l)$ are the neighbors of peer p at level l of the tree.

Algorithm 3: query(p, l_c, I_Q)

1 **if** $I_p \cap I_Q \neq \emptyset$ **then**
2 | return result tuples
3 **end**
4 $l_s = \max_l id_s \in I_{p,l}$;
5 $l_e = \max_l id_e \in I_{p,l}$;
6 $l_{min} = \max(l_c, \min(l_s, l_e))$;
7 $l_{max} = \max(l_s, l_e)$;
8 **if** $l_c < l_{max}$ **then**
9 | **for** $l = l_{min}$ **to** l_{max} **do**
10 | | select p' from $neighbor(p, l)$;
11 | | query($p', l_c + 1, I_Q$);
12 | **end**
13 **end**

In this algorithm, the peer first determines the levels at which the range covered by the subtree intersects with the query range (lines 4-5). This determines at which levels the search is forwarded to the peer's neighbors (line 10-11). However, if at an earlier stage of processing the query has already been forwarded at a given level, levels that are smaller than the current level of processing l_c need not to be considered (lines 6 and 8).

The use of structured overlay networks with multi-dimensional identifier space has been explored in various ways to support efficient range query processing. A one-dimensional attribute value domain can be mapped into an m-dimensional identifier space using a space-filling Hilbert curve [Andrzejak and Xu, 2002]. Since the Hilbert curve is locality-preserving, close attribute values

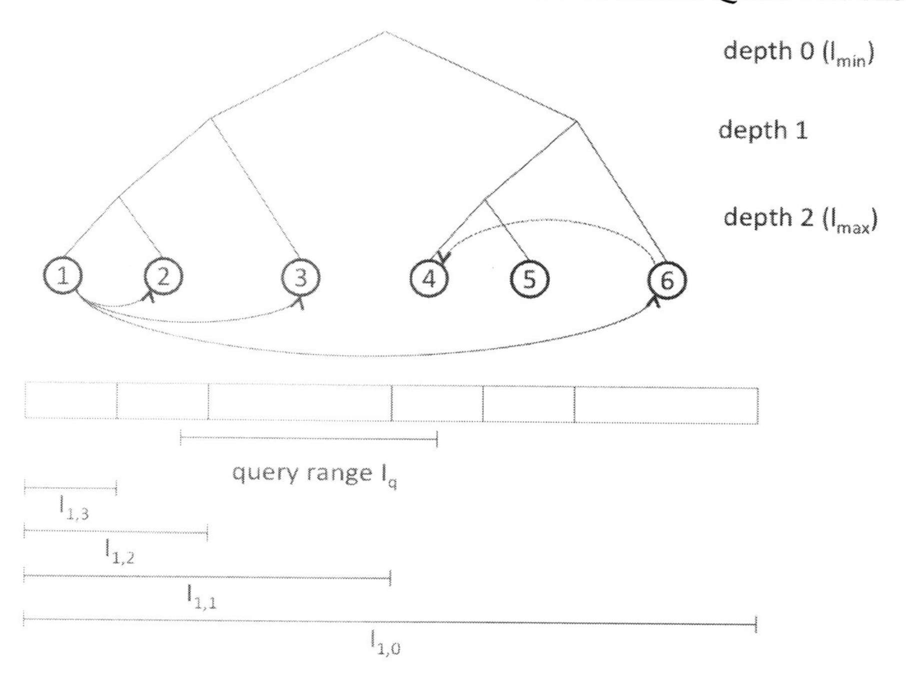

Figure 2.7: This figure illustrates the range query processing for range I_Q starting at peer 1. Since the ranges $I_{1,l}$ intersect at all levels l with I_Q, peer 1 will send a query message to each of its long range neighbors. Peer 6 will eventually forward the query to peer 4, which also intersects with I_Q.

will stay close in the multi-dimensional identifier space. This enables parallel search for query results and thus reduces latency. When a range query is processed, first the mid-point of the range is searched in the overlay network. Then the query is flooded in a controlled way to all neighbors along the different dimensions that hold a zone intersecting with the query range. The search process is illustrated in Figure 2.8.

Another use of structured overlay networks with multi-dimensional identifier space is to cache range query results in a 2-dimensional CAN overlay network [Sahin et al., 2004]. The two dimensions of the identifier space are used to represent the lower and upper bound of the query range. When a query Q_1 is subsumed by a query Q_2, of which the result has been cached earlier, i.e., $I_{Q_1} \subseteq I_{Q_2}$, the result can be obtained immediately from the cache stored in the overlay network.

2.2 COMPLEX QUERY PROCESSING

In practical applications, more complex queries than searches for single attribute values or attribute value ranges are required. In the following, we will introduce techniques that have been proposed for

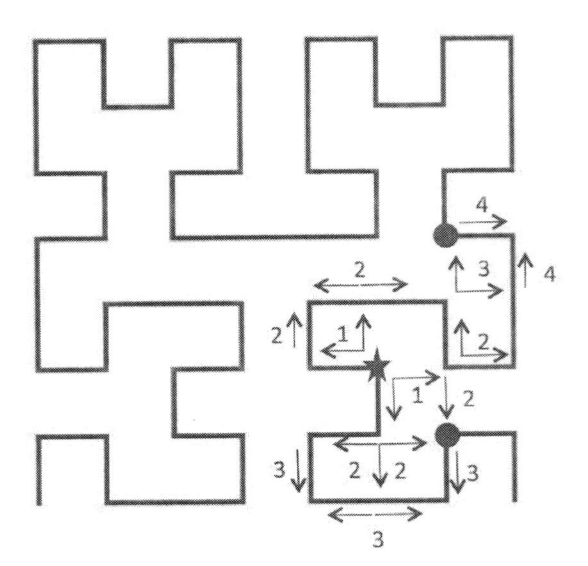

Figure 2.8: This figure illustrates the range query processing for range I_Q. The range mapped onto the Hilbert curve is indicated by the two bold points and the midpoint, where query evaluation is starting, is marked by a star. Note that after 4 steps, also, the last zone intersecting with the query is reached. Also some messages are flooded redundantly.

processing queries that are usually supported in relational database systems, in particular searches over multiple attributes, joins and aggregation queries.

2.2.1 MANAGING MULTIPLE ATTRIBUTES

Applications for resource management frequently search resources by combining conditions on multiple attributes (see Example 2.1). We study now the problem of supporting search over multiple attributes including range conditions. We assume that a fixed number of attributes $a_1, \ldots, a_k$ is given. We also assume that the number of attributes k is relatively small. Therefore, we are searching in a low-dimensional space. Generalizing the case of range queries on a one-dimensional attribute space, we will consider *conjunctive queries* over multiple attributes with equality or range conditions. For these types of queries, three categories of approaches for efficient processing using structured overlay networks can be identified.

1. Mapping attributes separately to a one-dimensional structured overlay network.
2. Mapping the k-dimensional attribute space into a one-dimensional space, and using a one-dimensional structured overlay network.

3. Using a structured overlay network with a k-dimensional identifier space and mapping attributes directly into this space.

Mapping to a single one-dimensional overlay network. Tuples $(v_1, \ldots, v_k)$ from a relation $R(a_1, \ldots, a_k)$ can be mapped to a one-dimensional overlay network by mapping each attribute value separately [Cai et al., 2004]. Each of the values v_j is normalized to the same range and then mapped by an order-preserving hash function to the peer identifier space I, e.g., in a one-dimensional Chord ring. The tuple is then stored at those peers which are responsible for one of the hashed values of $v_1, \ldots, v_k$. Each peer maintains for each attribute a_j a list of pairs $(v_j, (v_1, \ldots, v_k))$ for the tuples that it stores as a result of being responsible for the hashed value of v_j.

A naive approach to answer conjunctive range queries with this storage scheme, is to evaluate the condition for each attribute separately and then compute the intersection of the received results at the query originator. This is, however, in general, very costly and not necessary, since due to the storage scheme, the values of the other attributes are known, when looking up the value of a specific attribute a_j. Therefore, a more efficient approach is to first select the attribute for which the query condition is the most selective, and then iteratively traverse the range, while filtering out those tuples that do not satisfy the search condition. Note that this approach relies on the fact that every tuple is replicated k times in the overlay network.

A variation of this technique is to use a separate one-dimensional ring for every attribute [Bharambe et al., 2004]. As in the previous approach, attribute values are mapped to the corresponding peer and the peer stores the complete tuple together with the hashed value. In order to be able to resolve arbitrary queries, each peer maintains at least one link to a random peer in every other ring. Apart from that, search is performed analogously to the previous approach.

Mapping to a one-dimensional space An alternative approach to index multiple attributes in an overlay network, is to map the k-dimensional attribute space to a one-dimensional space using a *space-filling curve* [Ganesan et al., 2004b, Shu et al., 2005]. One example of such a space-filling curve is the *z-ordering* [Orenstein and Merrett, 1984]. The mapping is obtained by shuffling the bits of a binary representation of the attribute values. More precisely, if attribute a_j has a binary representation $b_{j1} \ldots b_{jl}$, then a tuple $(a_1, \ldots, a_k)$ is mapped to the bitstring $b_{11}b_{12} \ldots b_{1l}b_{21} \ldots b_{2l} \ldots b_{k1} \ldots b_{kl}$ of length $k \cdot l$. This mapping preserves locality as tuples with similar values in the multi-dimensional space will be mapped to close values in the one-dimensional space. Once the data have been mapped to the one-dimensional space, they are distributed over the peers using range-partitioning methods introduced earlier.

Processing range queries over multiple attributes is then performed in two steps. First the multi-dimensional query is transformed into a set of one-dimensional queries that is guaranteed to contain all query answers. The transformation algorithm recursively bi-partitions the multi-dimensional space till either the subspace is completely contained in the query range or it has an empty intersection [Orenstein and Merrett, 1984]. In a second phase each one-dimensional range query is evaluated as described earlier in Section 2.1.5.

Space partitioning schemes. In this approach the identifier space used for the overlay network has the same dimension k as the attribute space [Banaei-Kashani and Shahabi, 2004, Ganesan et al., 2004b, Schuett et al., 2008]. Multi-dimensional identifier spaces have for example been proposed for the CAN overlay network [Ratnasamy et al., 2001]. Different to CAN, where the space is always split into subspaces of equal size, for storing multi-dimensional attribute data, it is preferable to split the spaces into subspaces that carry equal load in terms of the number of tuples managed [Ganesan et al., 2004b, Schuett et al., 2008]. This corresponds to the approach that is taken in the construction of kd-trees [Bentley, 1975]. More precisely, when a new node arrives it searches the peer at which it wants to join, and splits the load with it evenly. The split dimensions are chosen cyclically. The peers maintain, as in CAN, links with their direct neighbors along all dimensions. For improving routing efficiency, long-range links can be added for each dimension following the same principle as described in Section 2.1.4 [Schuett et al., 2008]. The first long-range neighbor of a peer in dimension j is simply its neighboring peer along that dimension. To determine its $i + 1$st neighbor, a peer contacts the ith neighbor in its routing table for that dimension and requests its ith neighbor p. Then it sets its own $i + 1$st neighbor to this peer p. This process stops once the next neighbor in that dimension has an identifier larger than the peer itself. Depending on the data distribution this may lead to different numbers of neighbors along the different dimensions, but the total number will always be $O(\log n)$. The structure of such a multi-dimensional overlay network based on space partitioning is illustrated in Figure 2.9.

For processing multi-dimensional queries, first the request is routed to the peer holding the center of the query region. In contrast to the one-dimensional case in each routing step, there exist k alternative long-range links to choose from. Different strategies can be applied to choose among those using network proximity, distance or volume of the target cell as criteria. Space partitioning schemes also supports efficient processing of query regions other than hypercubes, such as hyperspheres.

2.2.2 JOIN QUERIES

So far, the most complex query types we have considered were conjunctive queries on multiple attributes. For supporting general relational query processing the missing element are join queries. Since data are horizontally fragmented in a peer-to-peer network, either by hash or range partitioning, the join algorithms are adaptations of corresponding algorithms from parallel and distributed databases [Huebsch et al., 2003].

In order to handle multiple relations, *namespace identifiers* are used. Assume a primary key a_j of a relation $R(a_1, \ldots, a_k)$ is indexed using a structured overlay network. Then a tuple is identified by $R.v_j$ where R is the relation name and v_j the value of the primary key attribute. The tuple is stored at the peer that is responsible for the identifier $h_R(R.v_j) \in I$. If the relation is indexed on non primary key attributes, then different tuples sharing the same key are distinguished by appending an additional instance identifier. For retrieving all tuples from a relation R on a peer p, a local scan operator $scan(p, R)$ is provided.

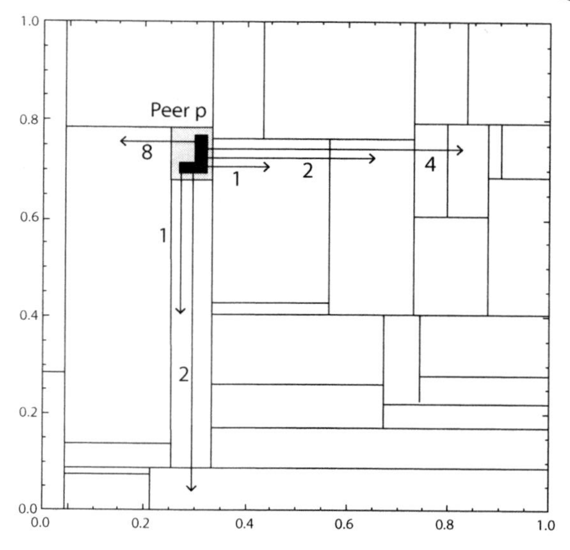

Figure 2.9: Example of a multi-dimensional overlay network using space partitioning of the two-dimensional attribute space and recursive construction of long-range links in the hop space (SONAR by [Schuett et al., 2008]). The figure shows the long range links of peer p along the two dimensions. The numbers next to the arrows indicate the distance in the hop space. Note that peer p has in addition 5 local neighbors it is connected to.

A first join algorithm is based on the *pipelining symmetric hash join*. For joining two relations R and S on attribute a_j, the following steps are performed:

1. The query is first broadcast to all peers, in order to locate those peers that hold tuples from R or S. Any broadcast algorithm for structured overlay networks can be used for that purpose [El-Ansary et al., 2003].

2. Using the scan operator, each peer retrieves the tuples of R or S that it stores. It inserts those tuples that satisfy the selection conditions of the join query into the peer-to-peer overlay network using a new temporary namespace T using the join attribute as key. In other words, a qualifying tuple t with attribute value v_j is inserted at the peer responsible for $h_R(T.v_j)$. Each tuple that is copied that way is additionally tagged with the name of the source relation.

3. Whenever a tuple from namespace T is received at a peer, a search for the value of the join attribute is performed on all tuples in T to retrieve all matching tuples from the other relation. This search is performed locally at the peer.

4. The pairs of tuples from R and S that match are sent to the query originator or to the node performing the next query processing step.

The processing is illustrated in Figure 2.10.

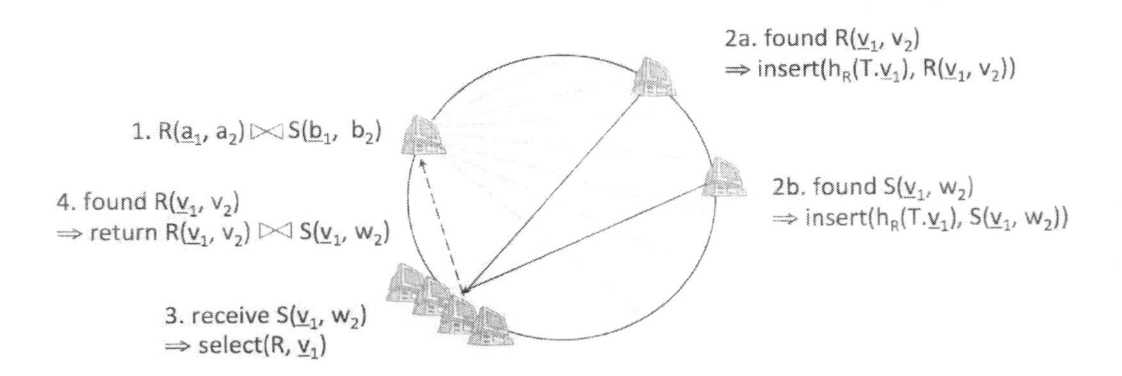

Figure 2.10: Illustration of a symmetric hash join. After broadcasting the query (step 1), two peers are found that hold tuples from the two relations to be joined. They insert those tuple in the namespace T using the join attribute value (step 2a and 2b). As a result, the tuples are inserted at the same peer. When a peer managing namespace T receives a tuple, it searches locally for matching tuples from the other relation (step 2). This search might span several neighboring peers. Finally, matching tuples are returned to the query originator (step 4).

In case one of the tables, say S, is already indexed on the join attribute, an alternative join algorithm, called *fetch join*, can be used [Huebsch et al., 2003]. The peers holding tuples from relation R retrieve all the tuples from R using a local scan. Then, for each tuple a search is performed for tuples from S with same join attribute value. These tuples are retrieved and sent to the query originator together with the tuple from R.

2.2.3 AGGREGATION QUERIES

Aggregation queries in peer-to-peer systems have significant interest from two perspectives. For system maintenance, aggregation queries play an important role in order to obtain global system statistics, such as the total number of peers or the workload distribution among peers. This aspect has already been partially covered in Section 2.1.4. From an application perspective, aggregation queries have been studied for supporting, for example, business decision support and resource allocation in grid computing.

The query types considered so far exhibit typically high selectivity and thus results can be retrieved from a few peers. They are therefore well supported by structured overlay networks. Aggregation queries typically concern a large number, or all peers, in a peer-to-peer network and can therefore as well be efficiently supported by unstructured overlay networks. Therefore, we will consider, in the following, techniques for both types of overlay networks

An aggregation query computes an *aggregation function* $\mu : V^n \rightarrow V$ over a domain of values V, where the function μ is defined for any $n \geq 1$. A typical situation that is frequently considered is the case of a peer-to-peer network with n peers, where every peer p_i contributes one value $v_i \in V$ to compute the aggregate value $\mu(v_1, \ldots, v_n)$. The value a peer contributes can be the result of some computation it performs on its local data, including the case of performing aggregation queries on its local database. The properties of the aggregation function μ essentially determine the potential to optimize processing of aggregation queries. We can distinguish three main classes of aggregation functions:

1. *Idempotent aggregation functions*: They satisfy $\mu(v, v) = \mu(v)$. The main examples for this type of aggregation function are computing a minimum or maximum value, min and max. Idempotency simplifies distributed processing of aggregation queries largely, since processing the same input values multiple times has no influence on the result.

2. *Associative aggregation functions*: They satisfy $\mu(\mu(v_1, v_2), v_3) = \mu(v_1, \mu(v_2, v_3)) = \mu(v_1, v_2, v_3)$. The main examples for this type of aggregation function are counting and summation, $count$ and sum. Associativity simplifies distributed processing of aggregation queries, since the order of processing the input values has no influence on the result.

3. *Algebraic aggregation functions*: These are aggregation functions that can be derived algebraically from the values of other aggregation functions, which may have additional properties, in particular associativity. Computing an average (avg) is an example of a non-associative aggregation function that can be derived from $count$ and sum, which are both associative.

4. Aggregation functions that do not exhibit the previous properties do exist, like computing a *median*. These are typically difficult to compute in a distributed peer-to-peer environment.

Another factor that strongly influences the design of techniques for processing aggregation queries are requirements on the availability of the aggregation result at a specific location or over time. With respect to availability at a location, we distinguish two basic cases. Either the query result needs to be known by a single peer, the *query originator*, or it needs to be known by all peers. A typical example for the first case is a business decision query, where one peer queries the whole network to extract some highly specific information. A typical example for the second case is system maintenance related statistics that is required by all peers, like the total number of peers in the system. The latter is also a typical example for the case of *continuous aggregation queries* where the aggregation query needs to be continuously evaluated over time; this in contrast to *ad-hoc aggregation queries*, which are evaluated only once and for which business decision queries are again an example.

Since the result of aggregation queries frequently provides statistical information, for many applications the exact result of an aggregation query is not required and a good approximation is sufficient. This opens the possibility of using randomized and sampling based techniques for increasing efficiency of processing while sacrificing result precision.

The basic objectives of efficient aggregation query processing are the minimization of communication cost, e.g., measured in numbers of messages exchanged among peers, and the fair balancing of the processing load among the peers. In addition, the quality of the result needs to be considered. This concerns two aspects. First, the presence of churn, where peers are constantly joining and leaving the system while processing an aggregation query, raises the question which state of the system the result corresponds to and what is a *valid* result for the query [Bawa et al., 2003]. Second, for approximate evaluation methods, the precision of the result, respectively, the error incurred, becomes an additional concern.

In the following, we will discuss representative methods that have been proposed for evaluating aggregation queries considering the above mentioned performance criteria. We first consider ad-hoc aggregation queries, then continuous aggregation queries, and finally conclude with sampling-based techniques for approximate evaluation.

Ad-hoc aggregation queries. We introduce, in the following, methods for processing ad-hoc aggregation queries in unstructured overlay networks [Bawa et al., 2003]. These methods extend, in a straightforward fashion, to structured overlay networks.

In an ad-hoc aggregation query, the result of the query needs to be known to the query originator only. A simple method is to broadcast the query to the whole network and let all nodes directly report their data back to the query originator. This, however, does not scale as the query originator would become a bottleneck. For aggregation functions that are associative, it is possible to distribute the work of result aggregation over all the peers in the network. This is in the simplest case achieved by constructing an *aggregation tree.* The construction proceeds in two phases:

1. Broadcast the query through the network. Each peer stores the peer from which it first received the query as its parent. Thus, a spanning tree is constructed in the overlay network with the query originator as root node.

2. Child nodes in the spanning tree are sending their data to their parents starting from the leaf nodes. Parent nodes wait till all child nodes have reported, aggregate the data, and then forward the aggregate result to their parents.

For an associative aggregation function, this algorithm will return the correct result to the query originator, and the workload is evenly distributed in the network. The problem with this algorithm in a dynamic peer-to-peer network is its sensitivity to failures of peers. A single node failing inside the aggregation tree can lead to wrong results, potentially loosing data from major parts of the network. In order to make the aggregation process robust, one can construct multiple aggregation trees. The following is an algorithm to construct k aggregation trees:

The query originator broadcasts a message containing the parameter k and a level information l. The query originator p_q sets its level information $l(p_q) = 0$, all other peers p initially set $l(p) = \infty$.

When a peer p_{child} receives the message from a peer p_{parent} with $l(p_{parent}) \neq \infty$ and its own level is $l(p_{child}) = \infty$, then it sets $l(p_{child}) = l(p_{parent}) + 1$, and forwards the query with the new level $l(p_{child})$ to all its neighbors. If the peer p_{child} had already an updated level information, it stores the level $l(p_{parent})$ of p_{parent}. The process terminates at a peer when the peer and all its neighbors have initialized their level information. Then the peers define their potential parents as

$$parents(p_{child}) = \{p \in N(p_{child}) \mid l(p) < l(p_{child})\} .$$

Every peer then uniformly randomly picks k parents from this set and in this way k aggregation trees are constructed. After aggregation trees are constructed, the aggregation function can be evaluated. Each peer waits until it receives the data to aggregate from all of its children. It aggregates this data and sends the results to all of its k parents. For idempotent aggregation functions, such as minimum and maximum, this will produce the correct result at the query originator.

For other aggregation functions, such as counting and summation, the same data will be transferred and included into the aggregation result multiple times. For obtaining correct results in such a setting, without excessively consuming memory, a space-efficient approximate aggregation method, such as *FM sketches* [Flajolet and Martin, 1983], can be employed. This method has been used for various problems of distributed and space-efficient aggregation, such as in data stream processing, sensor networks and peer-to-peer networks.

FM sketches. FM sketches count the number of different elements in a sequence S without explicitly storing the elements, as this set may be large.

We choose a parameter $l > \log_2(|I|)$ and initialize c bitvectors $\vec{b_i}$, $i = 1, \ldots, c$ of length l with zero. The size of c will determine the accuracy of the result. We assume that c independent random hash functions $h_i : I \rightarrow [0, l]$, $i = 1, \ldots, c$ are given. These hash functions have the property that they (approximately) map $n \cdot 2^{-(k+1)}$ elements to the value $k \in [0, l]$, i.e., half of the elements in I are mapped to 0, one quarter to 1, and so on.

For counting the sequence, S is sequentially scanned. For each element $p \in S$, the c hash values $h_i(p)$ are computed. Then the bit $h_i(p)$ in vector $\vec{b_i}$ is set to 1. Once all elements of S are scanned, for each vector the smallest bit z_i, which has the value 0, is identified. It can be shown that then the value

$$2^{\bar{z}}, \bar{z} = \frac{1}{c} \sum_{i=1}^{c} z_i$$

is an estimate with a bounded error of order $O(\frac{1}{c})$ (for details see Bawa et al. [2003]).

We describe how FM sketches are being used to perform space-efficient counting a set of peers $P_1 \subseteq P$ with multiple aggregation trees. When peer $p \in P_1$ contributes to the count, it generates the c hash values $h_i(p)$. To do that, it tosses up to l coins till the coin returns 1. The number of tosses gives the value k. It sets the k-th bit of a vector $\vec{b_i}$ to 1, all others 0. It then sends all bitvectors $\vec{b_1}, \ldots, \vec{b_c}$ generated this way to all parents. These aggregate the received bitvectors with their own bitvectors $\vec{b'_1}, \ldots, \vec{b'_c}$ by performing a logical OR operation for each vector, i.e.,

$$\vec{b'_i} := \vec{b'_i} \; OR \; \vec{b_i}, \; i = 1, \ldots, c \, .$$

The query initiator then eventually computes the final result from its own bitvectors. For computing a sum of integers, assume that the values to be added fall into an interval $[0, v_{max}]$. The size of the bitvectors is chosen as $l > \log_2(|I| \cdot v_{max})$. Then a peer contributing a value v to the sum increases the count v times. In a naive approach, this would require repeating the coin toss v times. But a simple argument shows that this is not required. The probability that bit b_j is 0 after v tosses is given as $\lambda = (1 - 1/2^j)^v$. So for each bit, one Bernoulli trial with probability λ is made to determine whether the bit stays 0.

For adding real numbers, this method is not adapted. To that end, an approach using extreme value statistics has been proposed [Baquero et al., 2009].

Continuous aggregation queries. In applications, such as Grid resource monitoring, aggregate values need to be computed continuously. The problem is then to compute at all times t at the query originator the value $\mu(t) = \mu(v_1(t), \ldots, v_n(t))$. Constructing an ad-hoc aggregation tree as described before, is not appropriate in this case. Rather, an aggregation tree needs to be continuously maintained while peers are leaving and joining the network. Maintaining such an aggregation tree is analogous to the problem of maintaining a structured overlay network. Therefore, it is natural to employ an existing structured overlay network as a substrate to construct the aggregation tree. In this way, the aggregation tree will be maintained as a side-effect of maintaining the overlay network.

One possibility to derive an aggregation tree for continuously computing aggregation queries is to use an existing Chord network [Cai and Hwang, 2007, Li et al., 2005]. In this approach, an identifier $q \in I$ is chosen that uniquely identifies the aggregation query, e.g., the hash value of the name of the attribute to be aggregated. The peer $p_q = successor(q)$ serves as the root of the aggregation tree. To build the aggregation tree rooted at peer p_q, each peer uses as its parent node the entry in its finger table that it would select to route a search for identifier q. Since Chord routes are loop-free and uniquely determined, this will result in a unique aggregation tree with depth $O(\log n)$. However, the tree will be unbalanced, as illustrated by Figure 2.11. Some peers will have many more children than others and thus carry a higher workload in the aggregation process.

In order to construct a balanced aggregation tree, peers choose as their parent nodes only entries in their finger table that are not farther away than a given distance. A precise distance bound can be given resulting in a balanced binary aggregation tree for the case of peers that are regularly distributed in the identifier space [Cai and Hwang, 2007]. A peer chooses its parent at a maximal distance of

$$2^{\lceil \log_2 \frac{4d_r + 2 \cdot d_m}{3} \rceil}$$

where d_r is the clockwise distance between the peer, and the root and d_m is the average distance between two adjacent peers. Figure 2.11 illustrates the effect of this distance bounding for a simple example.

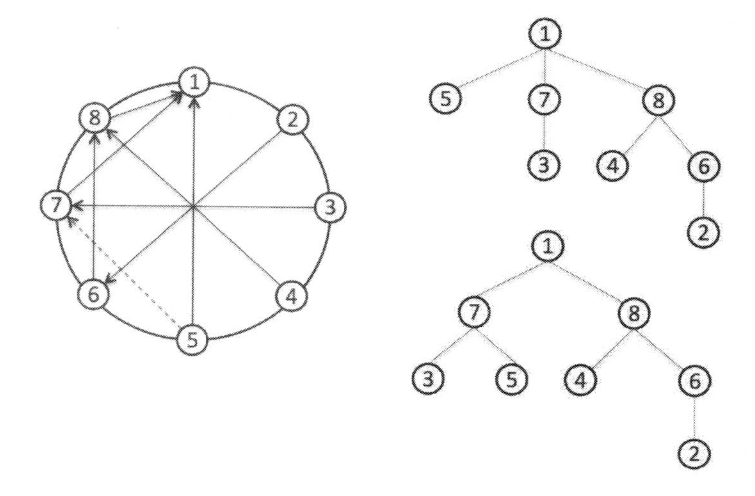

Figure 2.11: Construction of aggregation trees based on Chord. On the top we see the aggregation tree constructed from the shortest routes from a peer to the root. The resulting tree is unbalanced and not binary. When applying the distance restriction for choosing parent nodes, the peer 5 will be affected since $2^{\lceil \log_2 \frac{4+2}{3} \rceil} = 2$, and it has a direct link to the root at distance 4 that it is no longer permitted to use. Therefore, it follows the dashed link for routing towards the root.

For overlay networks that are based on the construction of distributed trees, such as Kademlia or P-Grid, the construction of a balanced aggregation tree is more direct [Albrecht et al., 2004].

When an aggregate value needs to be continuously known by all peers, e.g., the maximal load of a peer or the number of peers in the system, *epidemic protocols* are a suitable approach. The *anti-entropy protocol* is an epidemic protocol that has been originally designed to update replicas in distributed databases [Demers et al., 1987]. It can be applied to continuously aggregate the values of an attribute a [Jelasity and Montresor, 2004]. To that end, each peer p_i maintains an aggregate value a_i^{μ} and runs Algorithm 4.

In this algorithm, peers contact randomly chosen other peers and exchange with them their current aggregate values. Then both peers update their own aggregate value by aggregating it with the received value using the aggregation function μ. When the aggregation function is idempotent, e.g., minimum or maximum, every peer will eventually converge to the correct aggregate value, analogously as replicas are reconciled in distributed databases [Demers et al., 1987]. The algorithm is also suitable to compute the average of attribute values stored at the peers. For that purpose, the binary aggregation function $\mu(x, y) = \frac{x+y}{2}$ is used.

Sampling-based aggregation techniques. Since computing exact aggregate values can be too expensive, computing approximations of aggregate values has been considered as alternative. The basic

Algorithm 4: $update_{\mu}(p_1)$

```
1  repeat
2  |  wait(t_wait);                                  /* action of peer p1 */
3  |  peer p1 contacts a randomly chosen peer p2;
4  |  send a1^μ to p2;
5  |  receive a2^μ from p2;
6  |  a1^μ := μ(a1^μ, a2^μ);
7  until forever;
8  if peer p2 receives a1 then
9  |  send a2^μ to p1;                               /* action of peer p2 */
10 |  a2^μ := μ(a2^μ, a1^μ);
11 end
```

idea is to use an existing unstructured or structured overlay network to access a random sample of the peers, which provide sample data values from which the aggregate value is estimated. We can distinguish techniques that sample the set of peers and techniques that sample the local databases at the peers. In the first approach, communication cost is reduced since fewer peers need to be contacted, in the second approach, less data need to be transmitted to compute the aggregation query.

When sampling the set of peers, the main problem is to obtain an *unbiased sample* of peers, i.e., that every peer has the same probability of occurring in the sample. We consider the case of an unstructured overlay network and perform sampling through random walks. In a random walk, the next peer to be visited is uniformly randomly chosen among the peers to which the current peer is connected to. When the random walk is sufficiently long, the probability to visit a given peer will approach a stationary distribution. For a directed overlay network with a total of e edges, the probability of visiting a given peer p is $Pr(p) = \frac{deg(p)}{e}$, where $deg(p)$ is the in-degree of p. Thus, this distribution is in general non-uniform. If the graph is well-connected, this stationary distribution will be quickly reached [Gkantsidis et al., 2006].

We describe how aggregation using randomly sampled peers can be performed for evaluating a count query [Arai et al., 2006]. In this method, s peers are visited using a random walk. In order to more quickly approach a *stationary distribution*, only each j-th peer along the random walk is included into the set S of sampled peers, so that the total length of the random walk is $j \cdot s$. Each sampled peer $p \in S$ returns the local aggregate value $\mu(p)$ and its in-degree $deg(p)$. From these values, the query originator computes the probability $Pr(p)$ and an estimation of the global aggregate value

$$\mu_e = \frac{1}{s} \sum_{p \in S} \frac{\mu(p)}{Pr(p)}$$

which is an *unbiased estimator* for the true aggregate value, i.e., $E[\mu_e] = \mu(p_1, \ldots, p_n)$ where $E[\mu_e]$ is the expectation value of the random variable μ_e. The method can be further refined by estimating and controlling the sampling error. The query originator splits the set of sampled peers S randomly into two halves, S_1 and S_2, and computes a *cross-validation error*

$$\varepsilon_{cv} = |\frac{2}{s} \sum_{p \in S_1} \frac{\mu(p)}{Pr(p)} - \frac{2}{s} \sum_{p \in S_2} \frac{\mu(p)}{Pr(p)}| .$$

A simple statistical analysis [Arai et al., 2006] shows that the expected value of the square of the cross-validation error is two times the expected value of the square of the actual error, i.e., the variance.

$$E[\varepsilon_{cv}^2] = 2 \cdot E[(\mu_e - \mu(p_1, \ldots, p_n))^2] = Var[\mu_e]$$

It can also be shown that the variance is inversely proportional to the sample size. If the goal is to meet a minimal expected error ε_{min}, then additional samples can be taken depending on the cross-validation error. Based on this analysis, the number of the required additional samples is $s' = \frac{s}{2}\left(\frac{\varepsilon_{vc}}{\varepsilon_{min}}\right)^2$. The procedure of re-sampling and cross-validation can be repeated to increase the robustness of the method.

For computing more complex aggregation queries, in particular queries involving joins among several relational tables, methods for sampling relational databases that have been developed for *online aggregation* [Haas and Hellerstein, 1999, Hellerstein et al., 1997] can be adapted to the context of peer-to-peer data management. Evaluating these aggregation queries precisely would require the evaluation of relational joins, as described in Section 2.2.2. This might become prohibitively expensive. In many cases, it might also not be necessary if only approximate aggregation results are needed. As opposed to the previous sampling method, where a sample of peers aggregate all their local data and report it to the query originator, this approach relies on *data sampling*. All peers sample their local data and report the result back to the query originator. The query originator then performs the join and aggregation on the sampled data. This largely reduces communication cost, since only small samples of data are sent over the network [Wu et al., 2009].

In this approach, each peer p holding a subset R_p of a relation R provides a random sample of size $s \cdot \frac{R_p}{R}$ to the query originator in order to generate a sample of size s. Methods for obtaining such local samples, such as heap scans, index scans, and sampling via indices, have been described [Hellerstein et al., 1997]. Since the total size $|R|$ of the relation is not known initially the provisioning of samples is performed in two phases. In a first phase, all peers provide some samples and report their local relation sizes $|R_p|$. From this the query originator can determine the relation size $|R|$ and infer which peers have reported an insufficient number of samples. Those peers that have not reported a sufficiently large number of samples in the first phase provide the missing ones in a second phase.

The expected aggregate value can then be computed from the sampled data. In the case of a simple query involving a single relation such as

```
SELECT sum(op(t)) FROM R
```

the expected value is computed as

$$\mu_e = \frac{|R|}{s} \sum_{t \in S} op(t) \ ,$$

where the sample set is $S = \bigcup_p R_p$ with size $s = |S|$. Using a technique to estimate *large-sample confidence intervals* [Hellerstein et al., 1997], the error can then be estimated as

$$Pr[|\mu_e - \mu| \leq \varepsilon] \approx 2\Phi(\frac{\sqrt{s} \cdot \varepsilon}{\sigma_e}) - 1$$

where $\sigma_e = \sqrt{\frac{1}{s-1} \sum_{t \in S} (op(t) - \mu_e)^2}$ is an estimator for the variance of the values to be summed up. Φ is the cumulative distribution function of the standardized normal distribution.

In the case of an aggregation query involving multiple relations, the corresponding derivation of the error estimates becomes more complex. Consider an example of an aggregation query involving two relations such as

```
SELECT sum(op(t1,t2)) FROM R1,R2 WHERE t1.a = t2.b
```

The expected value is computed as

$$\mu_e = \frac{|R_1||R_2|}{|s_1||s_2|} \sum_{t_1 \in S_1, t_2 \in S_2, t1.a = t2.b} op(t_1, t_2) \ .$$

For determining an error bound, we need, as before, an estimator for the variance of the values being summed up. It is determined as follows. First, the marginal aggregate $\mu_{e,1}(t_1) = \frac{|R_1||R_2|}{s_2} \sum_{t_2 \in S_2, t1.a = t2.b} op(t_1, t_2)$ for $t_1 \in S_1$ is computed. It is easy to see that $\mu_e = \frac{1}{s_1} \sum_{t_1 \in S_1} \mu_{e,1}(t_1)$. Then, the marginal variance is $\sigma_{e,1}^2 = \frac{1}{s_1} \sum_{t_1 \in S_1} (\mu_{e,1}(t_1) - \mu_e)^2$. $\mu_{e,2}(t_2)$ and $\sigma_{e,2}^2$ are computed analogously. From this we obtain

$$\sigma_e = \frac{\sigma_{e,1}}{s_1} + \frac{\sigma_{e,2}}{s_2} \ .$$

From that, the error can be computed as in the case of one relation. Similar analysis can be performed for other operators such as count and average [Haas and Hellerstein, 1999, Hellerstein et al., 1997]. This method of computing an approximate aggregate for complex aggregation queries has been further refined in two ways [Wu et al., 2009]. First, by using multiple processors to compute the aggregation queries in parallel by partitioning the sample data. Second, by storing the sample sets as synopsis of the local databases for exploiting the samples for multiple queries.

Large-sample confidence intervals. We would like to estimate for a sample $S \subseteq R$ of size s how well the estimator $\mu_e = \frac{|R|}{s} \sum_{t \in S} op(t)$ approximates the true aggregate value $\mu = \sum_{t \in R} op(t)$. We assume that the sample size s is small enough such that the sample can be understood as a sample drawn with replacement but large enough such that the central limit theorem can be applied. Let $\sigma^2 = \frac{1}{|R|} \sum_{t \in R}(op(t) - \mu)^2$ be the variance of the values that are summed up. From the central limit theorem, it follows that

$$\frac{\sqrt{s} \cdot (\mu_e - \mu)}{\sigma}$$

is a random variable that that is normally distributed with mean 0 and variance 1. Now let $\sigma_e = \sqrt{\frac{1}{s-1} \sum_{t \in S}(op(t) - \mu_e)^2}$ be an estimator for σ (note that for $S = R$ we have $\sigma_e = \sigma$). Then also $\frac{\sqrt{s}(\mu_e - \mu)}{\sigma_e}$ is approximately normally distributed. Now we have that

$$Pr[|\mu_e - \mu| \leq \varepsilon] = Pr[|\frac{\sqrt{s} \cdot (\mu_e - \mu)}{\sigma_e}| \leq \frac{\sqrt{s} \cdot \varepsilon}{\sigma_e}] .$$

Since $\frac{\sqrt{s} \cdot (\mu_e - \mu)}{\sigma_e}$ is approximately normally distributed, we can conclude that

$$Pr[|\mu_e - \mu| \leq \varepsilon] \approx 2\Phi(\frac{\sqrt{s} \cdot \varepsilon}{\sigma_e}) - 1 .$$

Conclusion. Aggregation queries have received significant interest in peer-to-peer data management. The main strategies to aggregate data within a peer-to-peer network can be classified as follows:

- *Ad-hoc aggregation trees.* In this strategy, ad-hoc spanning trees are constructed, typically in a unstructured overlay network.
- *Overlayed aggregation trees.* In this strategy, a stable spanning tree is derived from a pre-existing structured overlay network. This tree is maintained as a side-effect of overlay network maintenance.
- *Epidemic protocols.* In this strategy, peers in a typically unstructured network continuously exchange their local views on the aggregate values which converge to the effective global aggregate value.

Solutions aim both at exact and approximate algorithms for answering this type of query. For approximate query answering, different statistical techniques are applied. We introduced the following techniques:

- *FM sketches.* Perform space-efficient approximate counting for large sets.
- *Peer sampling.* Select an unbiased sample of peers that provide their data to estimate the global aggregate value.
- *Data sampling.* Select an unbiased sample of data from the peers to estimate the global aggregate value for complex queries.

A key challenge in approximate methods is to determine reliable estimators for the approximation error, which we have illustrated for the different techniques. Given the wide variety of unexplored techniques for statistically analyzing large data sets and the wide range of applications for aggregation queries, this area is still wide open for further exploration. It is also closely linked to the recent developments in *peer-to-peer data mining* [Datta et al., 2006].

2.3 WEB DATA MANAGEMENT

The sharing and automated processing of data is the main objective of the Semantic Web. The Resource Description Framework (RDF) is a W3C standard providing a flexible and expressive data model for that purpose (http://www.w3.org/RDF/). Due to the inherently distributed nature of managing of RDF data on the Web, the question of applying peer-to-peer techniques for RDF data processing has received considerable attention.

Resource Description Framework (RDF). RDF is a semi-structured data model for modeling Web accessible resources and their relationships. Its basic construct is the *RDF triple*, consisting of a *subject*, a *predicate* and an *object*. Triples model binary relationships between resources or between a resource and its atomic properties. More precisely, triples are of the form $(s, p, o) \in (\mathcal{R} \cup \mathcal{B}) \times \mathcal{R} \times (\mathcal{R} \cup \mathcal{B} \cup \mathcal{L})$, where $\mathcal{R}$ are uniform resource identifiers, $\mathcal{B}$ are anonymous resources and $\mathcal{R}$ are literals, i.e., atomic data values. An example of triple would be $("http://www.morganclaypool.com/toc/dtm/1/1", editor, "Tamer Ozsu")$. In the XML encoding, this triple would be represented as

```
<rdf:Description rdf:about="http://www.morganclaypool.com/toc/dtm/1/1">
   <dc:editor>Tamer Ozsu</dc:editor>
</rdf:Description>
```

The RDF standard is complemented by *RDF Schema* (RDFS), a standard language to define schemas for RDF triples. The schemas can, among others, define classes of resources, types of predicates and constraints on the classes predicates can participate in. Several basic classes of queries have been identified and studied for RDF data.

- *Atomic queries.* Atomic queries constitute the basic building blocks for RDF queries and are expressed as *triple patterns*. They specify for each the subject, predicate and object either a constant value or a variable, such that a total of 8 different triple pattern query types exist. For example, the triple pattern $(?s, p, ?o)$ searches for triples with the predicate p matching any possible subject or object.

- *Range Queries.* Range queries are atomic queries where range conditions are added on some variables. For example, the query $(?s, p, ?v) \wedge ?v > c$ searches for subjects s that have for predicate p a value v larger than c.
- *Conjunctive queries.* Conjunctive queries consist of a conjunction of triple patterns or range queries, where variables can be shared among different conditions. For example, $?s : (?s, p_1, v1) \wedge (?s, p_2, v_2)$ searches for subjects by specifying two different conditions on its predicates. In general, we will denote a conjunctive query as $?x_1, \ldots, ?x_r : q_1, \ldots, q_k$, where $?x_1, \ldots, ?x_r$ are the result variables and $q_1, \ldots, q_k$ are triple patterns, possibly containing further variables.

For more complex queries, the SPARQL query language has been defined which provides expressivity comparable to SQL (http://www.w3.org/TR/rdf-sparql-query/).

Recently, RDF has become in particular popular through the Linked Open Data initiative (LOD) [Bizer, 2009], which consists of a rapidly growing collection of publicly accessible RDF repositories.

2.3.1 TRIPLE INDEXING

A basic approach to enable efficient access to RDF triples using structured overlay networks is to index the three components of a triple separately [Cai and Frank, 2004, Cudre-Mauroux et al., 2007, Liarou et al., 2006]. For a given triple, (s, p, o) the key value pairs $(h_R(s), (s, p, o))$, $(h_R(p), (s, p, o))$ and $(h_R(o), (s, p, o))$ are generated using an appropriate hash function h_r and inserted into the overlay network. For object domains that use literal values, order-preserving hash functions can be used in order to support range querying. The peers store the triples received that way in a local relational triple table T with the schema $T(s, p, o)$.

With this indexing technique, all atomic queries, except queries of the form $(?s, ?p, ?o)$, can be efficiently answered by routing to the peer responsible for one of the constant components of the triple patterns.

Conjunctive queries can be evaluated using the *query chain algorithm* [Liarou et al., 2006]. We describe the case of conjunctive queries without range conditions. A conjunctive query $Q = ?x_1, \ldots, ?x_r : Q_1, \ldots, Q_k$ is recursively resolved by processing it on a chain of peers $p_0, p_1, \ldots, p_k$, where each peer p_i processes the triple pattern Q_i and p_0 is the query originator.

At each processing step, the current peer p_i identifies a constant c in the next triple pattern Q_{i+1}, giving preference to subjects, over objects, and then over predicates. It then sends a query message $(Q, i+1, R_i, p_0)$ to peer p_{i+1} that is responsible for the constant c. Initially R is set to the empty relation $\{()\}$.

When a peer p_i receives a query message, it evaluates on its local triple table the query

$$R_{local} = \delta_{v_i \to x_j} \pi_{v_i} (\sigma_{Q_i^c}(T)) \ .$$

Q_i^c are the conditions implied by the constant part of the triple pattern Q_i, whereas v_i are the columns in the triple table corresponding to the variable part in the triple pattern. The variable renaming $v_i \to x_j$ maps the attribute names from the local triple table to the corresponding variable names of the query Q.

For example, if the triple pattern is $(?x_1, "author", "Aberer")$, then the query to compute R_{local} would be

$$\delta_{s \to x_1} (\pi_s (\sigma_{p="author" \wedge o="Aberer"}(T))) \ .$$

The peer then joins the local result with the current global result and projects on X_{i+1}, the set of variables consisting of the union of the answer variables $?x_1, \ldots, ?x_r$ and the variables in the remaining atomic queries $Q_{i+1}, \ldots, Q_k$:

$$R_{i+1} := \pi_{X_{i+1}} (R_i \bowtie R_{local})$$

If the query is not completely resolved, a message is sent to the next peer as described before, otherwise the result is returned to the query originator.

Query evaluation can be parallelized in order to distribute the query processing load among many peers [Liarou et al., 2006]. This requires to index for each triple (s, p, o) not only the individual components but also their combinations. Therefore, the triple is also inserted at peers responsible for the identifiers $h_R(s+p)$, $h_R(s+o)$, $h_R(p+o)$ and $h_R(s+p+o)$ where $+$ indicates the concatenation of a string representation of the values. When an intermediate result arrives, the next step of the query evaluation is dispatched according to the different matching values for the variables in the current triple pattern.

For example, if the query is $?x : (?x, p_1, v_1) \wedge (?x, p_2, v_2)$, and $S = \{s_1, \ldots, s_l\}$ is the set of subjects matching the first triple pattern, which can be found at the peer responsible for $h_R(p_1 + v_1)$, then the second triple pattern is evaluated by sending messages to the peers responsible for $h_R(s_1, p_2, v_2), \ldots, h_R(s_l, p_2, v_2)$. This method trades off better load balancing for higher indexing cost and increased message traffic during query processing.

When using order-preserving hash functions for indexing triples, as described above, not all possible combinations of triple components need to be indexed [Harth and Decker, 2005]. For example, the indexing of $h_R(s+p+o)$ supports not only the efficient lookup of constant triple patterns but it also supports efficient search for triple patterns of the form $(s, p, ?x)$ and $(s, ?x, ?y)$ by using range query processing. Thus, for example, the creation of identifiers of the form $h_R(o+s)$, $h_R(p+o)$ and $h_R(s+p+o)$, would suffice to answer directly all types of triple patterns. This eliminates the need to index the other combinations of triple components, i.e., indexing $h_R(s)$, $h_R(p)$ and $h_R(o)$. The query processing methods described above would be applicable under such an indexing strategy, by substituting requests to single peers by range queries processed over multiple peers.

The query processing strategies for conjunctive RDF queries can be further optimized by using caching techniques [Battre, 2008, Liarou et al., 2006].

2.3.2 SCHEMA-BASED INDEXING

When RDFS schemas are available for RDF data, they can be exploited for optimizing the data access [Nejdl et al., 2004]. Different peers might be specialized using specific schemas when storing data of a specific application domain. For this purpose, a superpeer architecture is particularly suitable. In such an architecture, the index information at a superpeer is organized in four levels:

1. *Schema index:* specifies schemas that a peer connected to the superpeer uses. Each peer is assumed to support one or more RDFS schemas, and the schemas are identified by a unique namespace identifier.

2. *Property index:* specifies properties that a peer connected to the superpeer uses. The properties are uniquely identified by the property name and schema identifier.

3. *Property value range index:* specifies ranges of values that properties at peers cover.

4. *Property value index:* specifies specific property values a peer uses.

Each superpeer maintains two indices of this type. One, called *superpeer-peer index,* maintaining the index information for all peers that are connected to the specific superpeer. One, called *superpeer-superpeer index,* maintaining index information for all other superpeers in the superpeer network. When a peer connects or disconnects to a superpeer, the local superpeer-peer index is updated correspondingly, and the update to the superpeer-superpeer index is broadcast in the superpeer network. For query processing, the characteristic elements of a query at the four index levels are extracted and used to route the search in the superpeer network to superpeers with matching index entries. These then forward the query to the corresponding peers.

Conclusions As RDF is becoming the de facto standard for Web data management, techniques for scalable and decentralized management of RDF data are gaining relevance. Though many of the techniques for relational data management in peer-to-peer settings are also of relevance to RDF data management, the use of RDF introduces novel qualities that need to be considered. On the one hand, this is the need to efficiently manage schema-less RDF data, a problem of semi-structured data management. On the other hand, this is the need to deal with large numbers of dynamically changing RDFS schemas that are continuously published in the Web.

CHAPTER 3

Peer-to-peer Data Integration

Integration of heterogeneous databases is one of the most prevalent problems in data management, in particular, in business and scientific applications. The basic problem addressed by *data integration* is the meaningful access to data from different databases that represent the same semantic concepts in syntactically different ways, i.e., overcoming *semantic heterogeneity*. This problem has been studied for different data models such as SQL or XML. *Data integration systems* provide the capability to correlate schemas expressed in the schema languages of these data models and to process queries over autonomous and heterogeneous databases.

The traditional approach to data integration is centralized. A common *global schema* for an application domain is defined, and mappings from the global schema to the specific application schemas are specified in order to overcome semantic heterogeneity. This process consists usually of the following two steps:

1. *Schema matching* is the identification of *schema correspondences* among schema elements that are considered as semantically related.

2. *Schema mapping* is the derivation of *schema mappings* that enable translations of queries and data that correspond to the different schemas and are considered as equivalent.

Tools for automated schema matching and mapping [Rahm and Bernstein, 2001] support these two tasks. After local application schemas are successfully mapped to the common global schemas, queries can be posed against the global schema and are automatically translated to a corresponding query that can be processed against the local application schemas. This approach is also called *view-based data integration*.

An alternative approach to overcoming semantic heterogeneity among different database systems is to support the *data exchange* among the systems [Fagin et al., 2003]. In contrast to view-based data integration, in data exchange it is assumed that a target system uses data from a source system and that the data from the source system is physically transformed and materialized at the target system. Again, schema mapping tools are used to generated the necessary mappings.

3.1 PEER-TO-PEER DATA INTEGRATION

Whereas data integration adopts a strongly centralized architectural paradigm, data exchange takes a purely localized view on the problem of integrating data from heterogeneous databases. From an architectural perspective, *peer-to-peer data integration* extends both of these traditional paradigms. It assumes a system environment where large numbers of peers, each with their proprietary schemas, mutually query or exchange data in the absence of central control and coordination. In the following, we provide an overview of the key ideas that have been developed for this system paradigm.

We describe a basic model of managing heterogeneous databases in a peer-to-peer setting. We assume a set of peers $P = \{p_1, \ldots, p_n\}$, where each peer p_i has a local schema S_{p_i} and a database instance $D(p_i)$ corresponding to this schema. The schemas and databases can be relational, XML or RDF/RDFS. Peers are connected by *mapping links* $m_{p_i \rightarrow p_j}$. A mapping link corresponds to the capability of translating queries and data among the databases that the two peers p_i and p_j hold. The mapping links form a directed graph that we can interpret as an unstructured peer-to-peer network. Each peer has only a small number of such mapping links to ensure scalability.

Example 3.1 Figure 3.1 illustrates a peer-to-peer network of databases related to today's Web ecosystem. The figure illustrates some salient features of peer-to-peer data integration. A peer-to-peer data integration system allows peers to access data from peers they are not directly connected to. For example, the peer *SocialNetworks,* can access *GoogleMail* by traversing the peer *Skype* and composing the mappings *SocialNetworks* → *Skype* and *Skype* → *GoogleMail.* We will call such compositions of mapping links also *mapping paths*. Some peers might be even reached along different paths, e.g., from $SocialNetworks$, there exist 8 different paths to reach $Twitter$.

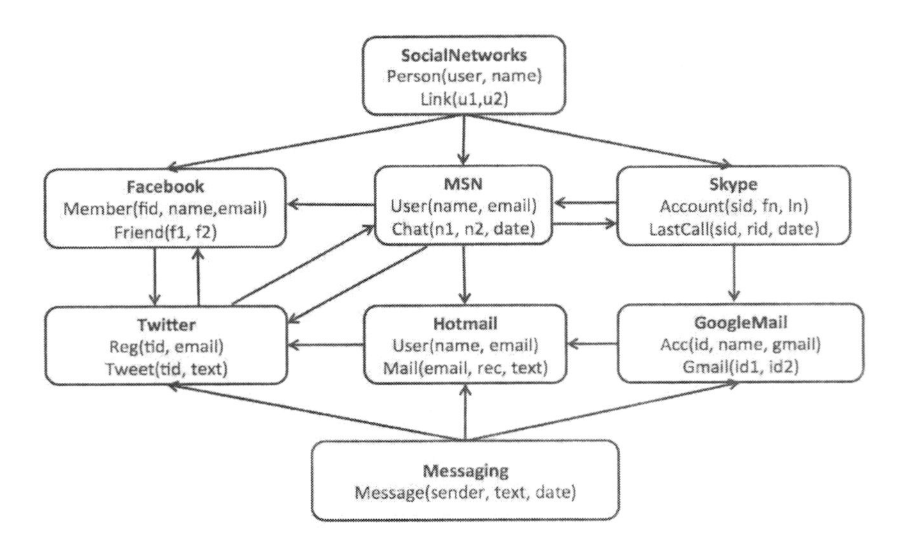

Figure 3.1: A peer-to-peer integration setting.

The main question we will explore in the following is how a peer can process a query against the peer-to-peer network by exploiting the network of mapping links.

3.2 QUERY REWRITING

3.2.1 ATTRIBUTE MAPPING

A basic approach to perform access to heterogeneous peer databases is by exploiting knowledge on schema correspondences in order to rewrite queries such that they match neighboring peers schemas and evaluate the rewritten queries at the neighboring peers. For example, in the relational model, these schema correspondences can identify corresponding relations and corresponding attributes within corresponding relations. The neighboring peers recursively translate the query for matching schemas of their neighbors till either no more translations are available or all available neighbors have already been accessed. The translation is, in the simplest case, performed by syntactically substituting the corresponding schema elements and producing results that are compatible with the schema of the query originator. The following algorithm describes the query processing strategy. The query originator p executes $resolveQuery(p, qid, Q, \{\})$.

Algorithm 5: $resolveQuery(p, qid, Q, P)$

1 $R = Q(D(p))$;
2 **for** $p' \in N(p) \setminus P$ **do**
3 $\quad Q' = m_{p->p'}(Q)$;
4 $\quad R_Q = resolveQuery(p', qid, Q', P \cup \{p\})$;
5 $\quad R = R \cup R_Q$;
6 **end**
7 return R;

In this algorithm, a peer first evaluates the query received from another peer against its database instance $D(p)$ (line 1). Then it forwards the query to all its neighbors that have not yet received the query (line 2). For that, it first translates the query using the available mapping to the neighbor (line 3). Q' is a query that can be executed against the schema of p' and returns results according to the result schema of the original query (line 4). Finally, the union of all results is sent back to the originator of the $resolveQuery$ message (line 7).

In the simplest case, correspondences are established by specifying *relation equivalences* and *attribute equivalences* [Aberer et al., 2002a, 2003, Cudre-Mauroux et al., 2007, Löser et al., 2003]. Depending on the data model, there exist different mechanisms of how these equivalences are specified. We illustrate the case of using RDFS, which is frequently used for knowledge representation in the context of Web data integration. In RDFS, the equivalence of classes and properties can be expressed using RDF statements.

Example 3.2 For the setting illustrated in Figure 3.1, the fact that the class *Person* in *SocialNetworks* corresponds to the class *Member* in *Facebook*, together with correspondences of attributes, could be expressed as follows:

```
<rdf:Description rdf:about="SocialNetworks:Person">
   <owl:equivalentClass rdf:ID="map" rdf:resource="Facebook:Member"/>
</rdf:Description>
```

Assuming that the structure of the two classes is the same, a SPARQL query such as

```
SELECT ?person WHERE (?person <rdf:type> <SocialNetworks:Person>)
```

can be rewritten to

```
SELECT ?person WHERE (?person <rdf:type> <Facebook:Member>)
```

The advantage of using RDF statements to specify equivalences in an RDF data management system, as described in Section 2.3, is that those statements can be stored and retrieved in the peer-to-peer RDF storage system like any other data [Cudre-Mauroux et al., 2007]. When a query is processed at a peer, after executing the query locally, it retrieves all possible equivalences of classes and properties occurring in the query. It translates the query by syntactic substitution, using the equivalences and sends the transformed query to the corresponding peers for execution as described in Algorithm 5.

3.2.2 MAPPING TABLES

Attribute mapping has limited expressibility for specifying schema mappings. Only one-to-one matching among relations and one-to-one mappings among attributes that belong to matching relations are supported. An approach for expressing more complex query rewritings for the relational data model is based on the concept of *mapping tables* [Kementsietsidis and Arenas, 2004].

A mapping table M relates the attributes of the relations of two peers, both at the schema and the instance level. This enables not only more complex schema mappings, including many-to-many mappings among relations, but also mapping of data instance mapping or as it is called, *entity matching* [Köpcke and Rahm, 2010].

Assume two peers p_1 and p_2 expose attributes $A = \{a_1, \ldots, a_{k_1}\}$ and $B = \{b_1, \ldots, b_{k_2}\}$, which may belong to different relations. A mapping table M is a relation over a subset of $A \cup B$. The values of the attributes in the mapping table can either be values from the original attribute domains or variables. Variables are used to establish an identity mapping among attributes, in the same way as we identified equivalent RDF properties using RDF statements in the previous section. Values are used to map data values that are semantically equivalent.

Example 3.3 We illustrate the concept of mapping tables by an example. Given peer p_1 with a relational table $User(nickname, firstname, lastname, lastchat)$ and a peer p_2 with the relational tables $Member(id, first, last)$ and $Conv(id, cid, start, end)$. Assume that correspondences between the two schemas are given by the following three mapping tables M_1, M_2 and M_3.

nickname	id
peer67	u123
peer67	u321
nopeer	u413

firstname	lastname	first	last
x	y	x	y

lastchat	start
z	z

The first mapping table M_1 specifies an attribute correspondence among the attributes *nickname* and *id*, together with a mapping of data values. M_2 and M_3 specify attribute correspondences with an identity mapping of the data values. For example, M_2 states that the attributes *firstname* and *first* are equivalent.

Note that many-to-many mappings among relations can be expressed by mapping tables. Assume that a relation of a peer p_1 that corresponds to two relations of another peer p_2 is given. Furthermore, assume the two relations of p_2 are connected by a common join attribute. Then the join attribute of the two relations occurs only once in the mapping table, and thus the tuples are correctly correlated.

Example 3.4 The three mapping tables of the previous example can be combined into one mapping table M as follows.

nickname	firstname	lastname	lastchat	id	first	last	start
peer67	x_1	y_1	z_1	u123	x_1	y_1	z_1
peer67	x_2	y_2	z_2	u321	x_2	y_2	z_2
nopeer	x_3	y_3	z_3	u413	x_3	y_3	z_3

This mapping table specifies a mapping from one table of peer p_1 to two tables of peer p_2. The join attribute *id* occurs only once in the mapping table, and thus it enforces the correct join of the tuples form relations *Member* and *Conv* of p_2.

A formal semantics of mapping tables and mapping table inference and composition can be given [Kementsietsidis et al., 2003]. For query processing, *query tableaux* can be used as a representation of select-join queries enabling an elegant formulation of the query translation step [Kementsietsidis and Arenas, 2004]. Given a select-join query of the form

$$\sigma_E(R_1 \bowtie \ldots \bowtie R_k) ,$$

the corresponding query tableau is constructed as follows. The set of attributes in the tableau is the union of all attributes in $R_1, \ldots, R_k$. The condition E is transformed to disjunctive normal form. Thus, E is a disjunction of conjunctions of the form $c_1 \wedge \ldots \wedge c_l$, and each condition c_i is either an equality condition $a_j = v$ on one attribute or a join condition $a_{j_1} = a_{j_2}$ on two attributes. It is assumed that attributes can occur only in a single condition. Then, for each conjunction, a row in the tableau query is generated as follows:

1. For an equality condition, the value v is inserted at attribute a_j.
2. For a join condition, a new variable x is inserted both at attributes a_{j1} and a_{j2}.
3. For attributes not occurring in the conjunction, a new variable x is inserted.

Example 3.5 The query Q_1

```
SELECT nickname
FROM user
WHERE nickname = firstname AND
      lastchat = "9am"
```

has the following tableau query representation Q_1^T.

nickname	firstname	lastchat
x	x	"9am"

In order to perform the translation of a query, we introduce the notion of *variable join* for relational tables containing variables. For two relational tables R_1 and R_2 with disjoint variable names we denote this operation by $R_1 \bowtie_{var} R_2$. Given two tuples $t_1 \in R_1$ and $t_2 \in R_2$, a result tuple is produced, depending on the values found in the common attributes of the two tuples:

1. If a common attribute has different constant values $c_1 \neq c_2$ in both tuples, then no result tuple is produced.
2. If a common attribute has the same constant value c in both tuples or a constant value c in one tuple and a variable x in the other, then the common attribute takes the value c in the result tuple.
3. If a common attribute has a variable in both tuples, then a new variable x is inserted for the common attribute in the result tuple.

Given a select-join query Q_1 at peer p_1 and a mapping table M providing a mapping to the schema of peer p_2, the query is rewritten as follows:

$$Q_2^T = \pi_B(Q_1^T \bowtie_{var} M) .$$

Q_1^T is the tableau query derived from Q_1 and B the set of attributes exposed by p_2. This translation is sound, i.e., consistent with the constraints imposed by the mapping table, and complete, i.e., resulting in the query producing the most general result [Kementsietsidis and Arenas, 2004].

Example 3.6 Computing the variable join for Q_1^T and M from example 3.4 results in query Q_2^T that is executed on peer p_2 and semantically equivalent to query Q_1^T. Note that, in this case, a query on a single table of p_1 is translated to a join query for p_2.

$$Q_2^T = Q_1^T \bowtie_{var} M \quad = \quad$$

id	*first*	*start*
x	x	"9am"

3.3 VIEW-BASED DATA INTEGRATION

Peer-to-peer data integration can be understood as a generalization of conventional data integration, where all data are mapped to a single global integrated schema. For conventional data integration, a powerful framework, based on first order logic, for characterizing and analyzing the possible mappings between global and local schemas has been developed. We will, in the following, discuss extensions of this framework that have been proposed for the case of peer-to-peer data integration.

3.3.1 GLOBAL-AS-VIEW MAPPINGS

The query rewriting approach to integrating data from heterogeneous peer databases can be described from the perspective of a single peer p as follows. Given a relation R_0 of the peer's schema, for each peer $p_i, i = 1, \ldots, n$ that is reachable through a mapping path, a query Q_i^R is generated that can be processed against the database instance $D(p_i)$ of p_i and produces results corresponding to the schema of R_0. From the perspective of the peer p, it constructs a global view R_{global} over the schemas of peers that are accessible in the peer-to-peer network by traversing a spanning tree of mapping links. The schema of R_{global} is compatible with the schema of R_0. This approach is known in the data integration literature as *global as view* or short as GAV approach [Lenzerini, 2002].

The semantics of this data integration approach can be given as the global view instance $D(R_{global})$:

$$D(R_{global}) = D(R_0) \cup \bigcup_{i=1,\ldots,n} Q_i^R(D(p_i)) \tag{3.1}$$

where $Q_i^R(D)$ is the result of evaluating Q_i^R over the database instance $D(p_i)$ and $D(R_0)$ is the local instance of relation R_0.

In the context of data integration, frequently, the *datalog* notation is adopted to describe data integration mappings [Gallaire and Minker, 1978, Lenzerini, 2002]. Applying datalog notation to the global-as-view data integration setting, we can write the mapping from equation 3.1 as follows

$$\begin{aligned} R_{global}(x_1, \ldots, x_k) &\;:-\; R(x_1, \ldots, x_k) \\ R_{global}(x_1, \ldots, x_k) &\;:-\; Q_i^R(x_1, \ldots, x_k), i = 1, \ldots, n \end{aligned}$$

In general, the mappings established through attribute correspondences or mapping tables as described in the previous sections can be expressed in datalog.

Datalog. Datalog is a database query language that enables recursive querying and is based on first order logic. It is a subset of the Prolog programming language. A *conjunctive query* Q in datalog is of the form

$$Q(\vec{x}) \; :- \; R_1(\vec{x}_1), \ldots, R_k(\vec{x}_k), C_1(\vec{y}_1), \ldots, C_l(\vec{y}_l)$$

where $R_1, \ldots, R_k$ are relation symbols and $C_1, \ldots, C_l$ are conditions using comparison operators. The variables in $\vec{x}, \vec{y}_1, \ldots, \vec{y}_l$ are all contained in $\vec{x}_1, \ldots, \vec{x}_k$. The expression at the left-hand side is called the head or goal of the query, the expression on the right-hand side the body. The semantics of a query in datalog for a database instance D can be given as a relational algebra expression.

$$Q(D) = \pi_{\vec{x}}(\sigma_{C_1(\vec{y}_1) \wedge \ldots \wedge C_l(\vec{y}_l)}(D(R_1) \bowtie \ldots \bowtie D(R_k))) \,,$$

where $D(R_i)$ is the instance of relation R_i. If multiple datalog rules are given for the same goal, the result of the query is the union of the results obtained for each rule. Recursive queries can be expressed by using the head of a rule in the body of another rule.

Example 3.7 The mapping specified by M in Example 3.4 in the previous section is represented in datalog notation as follows.

$User(nickname, firstname, lastname, lastchat) :-$
$Member(id, firstname, lastname), Conv(id, cid, lastchat, end), M_1(nickname, id)$

Using the additional expressivity datalog offers, the mapping could be refined by an additional condition, for example, as follows:

$User(nickname, firstname, lastname, lastchat) :-$
$Member(id, firstname, lastname), Conv(id, cid, lastchat, end),$
$M_1(nickname, id), end > lastchat$

A problematic aspect of the GAV approach as described so far, when applying it to a peer-to-peer setting, is that the query result may depend on the evaluation strategy taken in case the same peer could be reached through different mapping paths. Another problem is that different peers might obtain for semantically equivalent queries different results.

Example 3.8 Consider the example peer network shown in Figure 3.1. Assume that the peer $SocialNetworks$ has mapping links for the schemas of peers $Facebook$ and MSN

$$Person(x, name) \; :- \; Member(x, name, \_)$$
$$Person(x, name) \; :- \; User(x, name)$$

and the peer MSN has a mapping link for the schemas of peer $Facebook$.

$$User(x, name) \quad :- \quad Member(\_, x, \_)$$

Then a query such as `SELECT user FROM Person` would be translated to `SELECT fid FROM Member` when using the direct link among $SocialNetworks$ and $FaceBook$, and to `SELECT name FROM Member` when passing through MSN. Thus, a confusion of the attributes takes place, and the same original query mapped to the same schema produced different translated queries. The query result would thus depend on which mapping path is chosen.

3.3.2 GLOBAL-AND-LOCAL AS VIEW MAPPINGS

Applying the GAV integration paradigm for peer-to-peer data integration requires query semantics that is independent of the query evaluation strategy. In the following, we will introduce an approach that achieves this and extends the expressivity of mappings beyond GAV mappings [Halevy et al., 2003]. In this approach, the mappings among the peers are considered as globally valid constraints of a distributed relational database.

In addition to the GAV type of mappings, this approach supports *conjunctive inclusion dependencies* (Koch [2004]). A conjunctive inclusion dependency is, in general, of the form $Q_1(\vec{x}) \subseteq Q_2(\vec{y})$ where Q_1 and Q_2 are conjunctive queries. The semantics of an inclusion dependency is given as

$$\{\vec{z} \mid \exists \vec{x}_{-z} Q_1(\vec{x})\} \subseteq \{\vec{z} \mid \exists \vec{y}_{-z} Q_2(\vec{y})\},$$

where $\vec{z}$ are the common variables in the two queries, and $\vec{x}_{-z}$ and $\vec{y}_{-z}$ are the variables without $\vec{z}$. We will also write in short $Q_1 \subseteq Q_2$ in the following. If $Q_1 \subseteq Q_2$ and $Q_2 \subseteq Q_1$, then we write $Q_1 = Q_2$. If the left-hand side of an inclusion dependency is a relation symbol, i.e., the inclusion dependency is of the form $R \subseteq Q$, then it specifies a *local-as-view* mapping or short LAV mapping [Lenzerini, 2002]. Answering queries with LAV mappings requires techniques for query answering using views [Abiteboul and Duschka, 1998]. In the general case, when both sides of an inclusion dependency are non-trivial queries, the corresponding mapping is called a *global-and-local as view* or GLAV mapping.

A model for a peer-to-peer data integration system using inclusion dependencies to specify peer-to-peer schema mappings is then given as follows. Each peer p has its relational *peer schema* S_p that it exposes to other peers. In addition, a peer can have local *stored relations* R, not necessarily contained in the peer schema, that are related to its peer schema by a *storage description*. The mapping among a stored relation R and the peer schema S_p is defined by an inclusion dependency $R = Q$ or $R \subseteq Q$. The intended meaning is that the data of R equals or is contained in the result of evaluating Q over the peer schema S_p.

Different peers can relate their schemas by *peer mappings* of the form $Q_1 = Q_2$ or $Q_1 \subseteq Q_2$, where Q_1 and Q_2 are conjunctive queries over arbitrary subsets of peer schemas or stored relations at peers. This type of mapping supports global-and-local as view or GLAV mappings among peer schemas.

A second type of peer mappings is *definitional mappings* for expressing GAV mappings. They are of the form $R :- R_1, \ldots, R_k$ where the relation symbols R and $R_1, \ldots, R_k$ are peer relations from the peer schemas. Multiple such definitional mappings for the same relation R express that R is the union of the results of all queries found on the right-hand side. Note the different meaning of specifying the definitional mappings $R :- R_1$ and $R :- R_2$ and the peer mappings $R = R_1$ and $R = R_2$. The latter implies that $R_1 = R_2$, whereas the former expresses that R is the union of R_1 and R_2.

An example of such a peer-to-peer integration system is given in Figure 3.2. It contains definitional mappings (e.g., from peer 1 to peer 2), peer mappings (e.g., from peer 1 to peers 3 and 4) and storage mappings (for peers 2, 3 and 4).

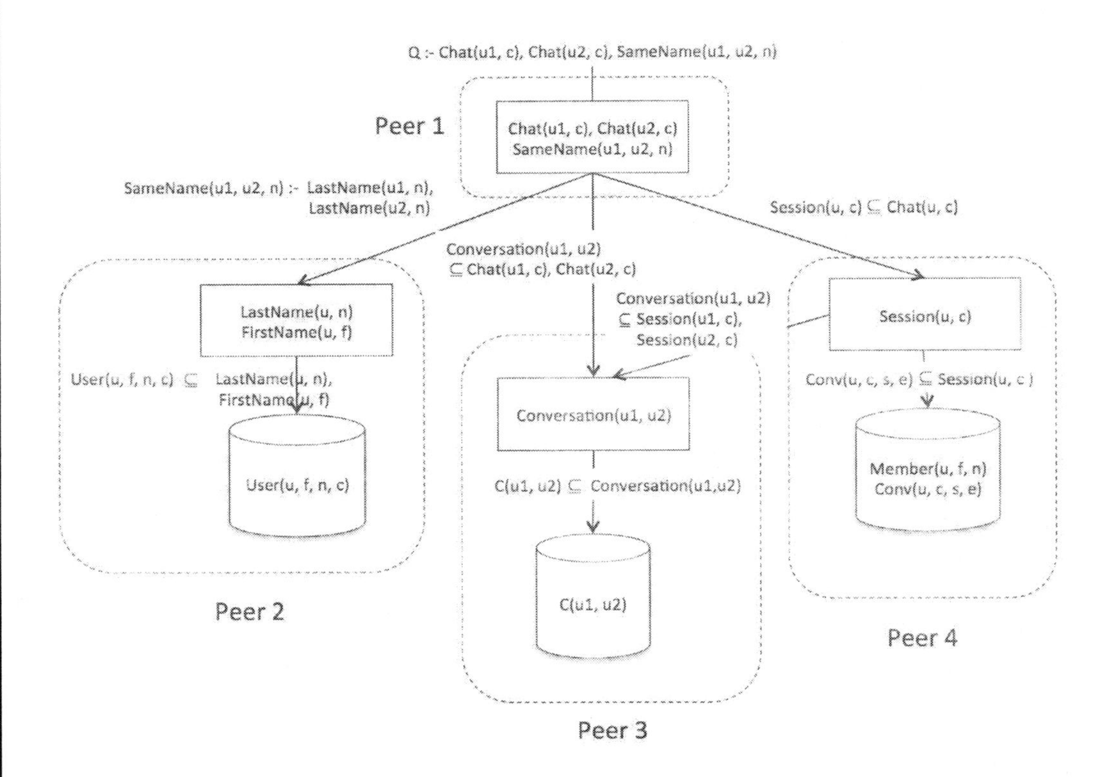

Figure 3.2: A peer-to-peer integration setting.

For this model, a global semantics for query answering can be given [Halevy et al., 2003]. A *storage instance* D assigns a set of tuples $D(R)$ to all stored relations R of peers. For processing queries over the peer schemas, which are views over the stored relations, also instances for the relations in the peer schemas are needed. A *data instance* D_P for the peer-to-peer system is an assignment of sets of tuples $D_P(R)$ for all stored and peer relations R. $Q(D_P)$ denotes the result of evaluating a query Q over a data instance D_P. For the same storage instance for the stored relations, multiple data instances for the peer-to-peer system might exist.

Example 3.9 Assume that the following LAV mappings are given, where $User$ and $Chat$ are stored relations.

$$\begin{aligned} User(x) &:- \; Session(x, c) \\ Chat(c) &:- \; Session(x, c) \end{aligned}$$

A storage instance D of the stored relations could be $D(User) = \{(Tamer)\}$, $D(Chat) = \{(7)\}$. Different data instances D_P are now possible for the peer relation $Session$, such as $D_1(Session) = \{(Tamer, 7)\}$ or $D_2(Session) = \{(Tamer, 6), (Karl, 7)\}$.

Therefore, we need to characterize which data instances are consistent with a storage instance: a data instance D_P is consistent with a storage instance D if

- for every storage description $R = Q$ or $R \subseteq Q$, we have $D(R) = Q(D_P)$, respectively, $D(R) \subseteq Q(D_P)$
- for every peer description $Q_1 = Q_2$ or $Q_1 \subseteq Q_2$, we have $Q_1(D_P) = Q_2(D_P)$, respectively, $Q_1(D_P) \subseteq Q_2(D_P)$
- for every definitional description $R :- Q_i, i = 1, \ldots, k$, we have $D(R) = Q_1(D_P) \cup \ldots \cup Q_k(D_P)$

Since multiple data instances D_P might be consistent with an instance D for the stored relations, for query answering, only result tuples are accepted that would be produced in every consistent data instance. This leads to the notion of *certain query answer*.

Let Q be a query over the schema of a peer p and let D be a storage instance. A tuple t is a *certain answer* to Q if $t \in Q(D_P)$ for every data instance D_P that is consistent with D. We denote the certain answer to Q given D by $Q(D)$.

Example 3.10 Consider the situation described in Example 3.9 and the query $Q(x, c) :- Session(x, c)$. Then we obtain as result $Q(D) = \emptyset$, since no tuple is an answer for all possible data instances. So there exists no certain answer.

It turns out that under this model query answering is, in general, not decidable, even not for conjunctive queries. Therefore, restricted classes of mappings are considered. For example, if storage descriptions and peer mappings are only of the inclusion type and all mappings are acyclic, then conjunctive queries can be answered in polynomial time. Excluding equality for peer mappings completely is, however, too restrictive for practical purposes. Equality might be required to express that two relations at different peers are replicas of each other. Therefore, another result shows that query answering is also polynomial under the following restrictions:

1. Whenever a storage description or peer mapping is of equality type, it does not contain projections.

2. A peer relation that appears in the head of a definitional mapping does not appear on the right-hand side of any other mapping.

Query processing, in this setting, requires transforming a query Q over the peer schemas into a query Q' that refers only to stored relations of peers. The query processing algorithm consists of two main steps [Halevy et al., 2003]. For definitional mappings, a standard view expansion is performed, as described earlier for the GAV case. For peer mappings and storage descriptions, techniques for query answering using views need to be applied. The basic principle of query answering using views works as follows. Given a peer mapping $Q_1 \subseteq Q_2$ and a query of the form $Q :- Q_2, Q'$, the query can be rewritten as $Q :- Q_1, Q'$. All possible rewritings of this kind are applied, and the result of the query consists of the union of the results of all rewritten queries.

We provide in the following an example of processing a query that illustrates the main issue with view expansion of LAV mappings.

Example 3.11 Assume that we want to evaluate the query

$$Q(u_1, u_2) :- Chat(u_1, c), Chat(u_2, c), SameName(u_1, u_2, n)$$

for the peer-to-peer network shown in Figure 3.2. For the $SameName(u_1, u_2, n)$ relation, we have a definitional mapping from peer p_1 to peer p_2. Applying this mapping, we rewrite the query

$$Q(u_1, u_2) :- Chat(u_1, c), Chat(u_2, c), LastName(u_1, n), LastName(u_2, n)$$

For the $Chat(u, c)$ relation, we have two peer mappings, which are LAV mappings.

$$\begin{aligned} Session(u, c) &\subseteq Chat(u, c) \\ Conversation(u_1, u_2) &\subseteq Chat(u_1, c), Chat(u_2, c) \end{aligned}$$

Both can be used to rewrite the right-hand side of the query. Furthermore, the second mapping can be applied in two different ways since the relation $Conversation(u_1, u_2, n)$ is not symmetric in the variables u_1 and u_2. Therefore, we obtain

$$\begin{aligned} Q(u_1, u_2) &:- Session(u_1, c), Session(u_2, c), LastName(u_1, n), LastName(u_2, n) \\ Q(u_1, u_2) &:- Conversation(u_1, u_2), LastName(u_1, n), LastName(u_2, n) \\ Q(u_1, u_2) &:- Conversation(u_2, u_1), LastName(u_1, n), LastName(u_2, n) \end{aligned}$$

At this stage, also the mapping

$$Conversation(u_1, u_2) \subseteq Session(u_1, c), Session(u_2, c)$$

from peer p_4 to peer p_3 could be applied. However, it would result in a rewritten query that is already subsumed by a query obtained so far, so this rewriting is redundant. In a final step, all the storage mappings are applied, resulting in a query to be executed against the stored relations at the peers.

$$
\begin{aligned}
Q(u_1, u_2) &\;:-\; Conv(u_1, c, s_1, e_1), Conv(u_2, c, s_2, e_2), \\
&\quad\;\; User(u_1, f_1, n, c_1), User(u_2, f_2, n, c_2) \\
Q(u_1, u_2) &\;:-\; C(u_1, u_2), User(u_1, f_1, n, c_1), User(u_2, f_2, n, c_2) \\
Q(u_1, u_2) &\;:-\; C(u_2, u_1), User(u_1, f_1, n, c_1), User(u_2, f_2, n, c_2)
\end{aligned}
$$

A comprehensive description of query processing algorithms for view-based integration in a peer-to-peer setting is found in the works of Halevy et al. [2003], Pottinger and Halevy [2001] and Koch [2004].

From a theoretical perspective, there have been attempts to overcome the limitations on the expressivity required for obtaining tractable query processing. The problems of this approach stem largely from the fact that the peer-to-peer database is considered as one single logical database with no separation of concerns for the contributions of the autonomous peers. One approach by Calvanese et al. [2004a] interprets peer-to-peer data integration as a *data exchange problem* [Fagin et al., 2003]. Another approach by Calvanese et al. [2004b] is based on epistemic logics, where constraints introduced by peers are not considered as global truths, but as local knowledge of peers.

3.4 INCOMPLETE AND INCORRECT MAPPINGS

Mapping links are established by peers either manually or by using schema matching and mapping tools [Rahm and Bernstein, 2001]. Establishing mapping links is susceptible to imperfection. On the one hand, mappings might be *incomplete* and not be able to map all parts of a query to a form that may be processed against the schema of another peer, even if that peer database contains information that pertains to the semantics of the query. On the other hand, mappings might be *incorrect* producing queries that retrieve results that do not correspond to the intended semantics of the query, as illustrated earlier in Example 3.8.

In fact, avoiding such undesirable situations is one of the reasons why the query resolution algorithm 7 avoids potential semantic inconsistencies proactively by not admitting query translation paths that are redundant. Such inconsistencies are likely to occur in large scale peer-to-peer data management systems with autonomous peers and thus pose a problem. But at the same time, they also imply an opportunity: inconsistencies may be considered as an additional information source produced by the peer-to-peer network, helping to evaluate the quality of existing mappings.

We will now present methods for dealing both with incompleteness and incorrectness of mapping links.

3.4.1 HANDLING INCOMPLETE MAPPINGS

When attempting to map a query in order to match the schema of another peer, it might occur that not for all parts of the query such a mapping is available. Typically, there might be attributes for which no mapping exists. Then the peer could either refrain from translating the query or relax the query by dropping parts of the query. We investigate now the relaxation approach for the case of GAV mappings [Aberer et al., 2003].

We consider simple select-project-join queries. When an attribute occurring in the projection cannot be mapped, it is dropped from the query. The query originator will not receive values for this attribute. Similarly, when an attribute in a condition cannot be mapped, the condition is dropped from the query, and the condition will not be evaluated. As a result, the peer might receive undesired results.

Relaxation of queries degrades thus the quality of results, and at a certain point, may render the mapped query useless. Therefore, a measure is introduced to assess the quality of the relaxed query. The query is only forwarded to other peers if the quality is above a certain threshold. Let $A(Q)$ be the set of attributes in the projection of a query Q, and $C(Q)$ the set of conditions occurring in the selection. For an original query Q that is translated to a query Q_m, the following similarity measures are defined to measure the completeness of translation:

- *Projection similarity*: $sim_A(Q, Q_m) = \frac{\vec{w} \cdot \vec{\delta}(A(Q))}{|\vec{w}|_2 |\vec{\delta}(A(Q))|_2}$ where $\vec{\delta}_i(A(Q)) = 1$ if attribute $a_i(Q)$ could be mapped to an attribute in Q_m and w_i is the relative importance of attribute $a_i(Q)$ for the original query.

- *Selection similarity*: $sim_C(Q, Q_m) = \frac{\vec{w} \cdot \vec{\delta}(C(Q))}{|\vec{w}|_2 |\vec{\delta}(C(Q))|_2}$ where $\vec{\delta}_i(C(Q)) = selectivity(C_i(Q))$ if condition $c_i(Q)$ could not be mapped to a condition in Q_m and $\vec{\delta}_i(C(Q)) = 1$, otherwise. $selectivity(C_i(Q))$ is the selectivity of the condition for the relation to which it is applied. w_i is the relative importance of condition $c_i(Q)$ for the original query.

Only if both similarity values are above a given threshold, the query will be forwarded. This approach can be generalized in different directions [Kantere et al., 2009], in particular, it is applicable also when GLAV mappings are used. Furthermore, peers can compare mappings composed along different paths, using similarity measures for queries and decide to drop certain mapping links if alternative paths produce better results. In this way, the peer-to-peer data integration network is effectively restructured.

3.4.2 DISCOVERING MAPPINGS

Usually, it is assumed that mapping links in peer-to-peer networks are established locally at a peer, either manually or by using schema mapping tools. These mappings are derived from features of the two schemas for which the mapping is established. A peer-to-peer network might, however, offer other information sources to support the process of establishing schema mappings. Different methods exist for that purpose.

Schema annotations. It is possible to annotate relations and attributes of the schemas at a peer with keywords in order to enable the identification of semantically related schemas [Ng et al., 2003]. Each relation R and attribute a is annotated at peer p with a set of keywords $K_p(R)$, respectively, $K_p(a)$. When evaluating a query Q, the peer p determines the set of relation and attribute annotations for the schema elements occurring in the query. Let $K_R(Q) = \bigcup_{R \in Q} K_p(R)$ and $K_A(Q) = \bigcup_{a \in Q} K_p(a)$ be those annotations. Then p searches for other peers p' that have related schemas by using the following similarity measure between the query and relations R at other peers:

$$sim_Q(Q, R) = \frac{w_R \cdot sgn(|K_{p'}(R) \cap K_R(Q)|) + w_A \cdot |\bigcup_{a \in R} K_{p'}(a) \cap K_A(Q)|}{w_R + w_A \cdot |K_A(Q)|}$$

$sgn(|K_{p'}(R) \cap K_R(Q)|)$ is 1 if the relations in Q and R have some common keyword and zero, otherwise. $|\bigcup_{a \in R} K_{p'}(a) \cap K_A(Q)|$ counts the number of keywords of attributes in Q, matching keywords of attributes in R. w_R and w_A indicate the relative importance given to the relation and attribute matching. $|K_A(Q)|$ is the total number of keywords found for attributes in Q. This retrieval method is used to propose to a user of the system potentially matching relations and a reformulated query that the user can interactively select for further processing.

Database Probing. Another approach to search for relations related to a query is by probing the databases stored at the peers [Li et al., 2007]. In this approach, it is assumed that a classification hierarchy is given. It consists of classes C and their attributes. Furthermore, classification rules are available that classify relations with respect to the classification hierarchy. The conditions of these rules are based on the data that is found in the relations: a classification rule is of the form $K \rightarrow C$ where K is a set of keywords and C a class in the hierarchy. Given such a classification rule, a peer can generate a *query probe* that retrieves in the target relation the number of tuples that contain in their attribute values all the keywords in the head of the classification rule. $W(R, C)$ counts the total number of results obtained for all classification rules for C. Then the confidence of a relation, belonging to a given class, is determined for non-root classes C by

$$sim_P(R, C) = sim_P(R, parent(C)) \cdot \frac{W(R, C)}{\sum_{C' \in children(parent(C))} W(R, C')}$$

For the root class C_r of the classification hierarchy $sim_P(R, C_r) = 1$. At each level, this measure gives the highest similarity to the child class that has received the highest relative number of matching keywords. As one descends the classification hierarchy, the similarity gradually diminishes. At some point, for all children C' of a class C either the similarity $sim_P(R, C')$ or the keyword count $W(R, C')$ drops below a given threshold. The relation is then associated with that class C. Thus, the classification can proceed in a top down fashion. A relation is first inserted at the root and then pushed down the hierarchy till the threshold conditions are no longer met. Once a matching relation has been found, the single attributes can be matched, using known constraints on the value domains and again database probing.

3.4.3 CORRECTING MAPPINGS

A distinctive feature of peer-to-peer data integration is the possibility to transitively compose mappings over multiple mapping links to *mapping paths*. This implies that the same node may be reached through different compositions of mapping links. If the query originator is reached by a mapping path, we have a *mapping cycle*. Figure 3.3 illustrates such situations. For example, in Figure 3.3(c), we see two parallel mapping paths from peer p_1 to p_2 and two cycles of mappings starting from peer p_1. The standard query evaluation strategy, described in Algorithm 7, avoids, in such a situation, redundant query evaluation by not traversing peers that have already reached by another mapping path. This also avoids potential problems with inconsistent mappings along different mapping paths. However, redundant mapping paths are not only a source of potential inconsistencies but they also constitute a potential source of information that can be exploited to assess the quality of mappings.

In order to systematically evaluate the quality of mappings by checking consistency of redundant mapping paths, a probabilistic framework can be applied [Cudre-Mauroux et al., 2006]. Assume an attribute a is mapped along two redundant paths $M_1 = m_1, \ldots, m_{k_1-1}$ and $M_2 = m_{k_1}, \ldots, m_k$ to attributes a_1 and a_2. We can consider cycles as a special case of redundant paths, where one path is of length zero and $a_1 = a$. Let f be the result of testing whether a_1 equals a_2. Then we can define the following probabilistic model for this observation to occur:

$$Pr(f = 1|m_1, \ldots, m_k) = \begin{cases} 1, & \text{if } m_1, \ldots, m_k \text{ are correct} \\ 0, & \text{if exactly one element in } m_1, \ldots, m_k \text{ is incorrect} \\ \Delta, & \text{if several elements in } m_1, \ldots, m_k \text{ are incorrect} \end{cases}$$

The quantity Δ characterizes the likelihood that multiple mapping errors compensate each other. This quantity could be derived either from system-wide statistics or by considering features of the involved mappings. Given this model, we would like to infer from such observations f the probability that a specific mapping m is correct. In order to do this, we apply Bayes theorem. For example, if we would like to know whether the mapping m_1 is correct, denoted as $m_1 = 1$, given that $f = 1$, we can write

$$Pr(m_1 = 1|f = 1) = \sum_{m_1=1, m_2, \ldots, m_k} \frac{Pr(m_1) \ldots Pr(m_k) Pr(f = 1|m_1, \ldots, m_k)}{Pr(f = 1)}$$

where $m_i, i \neq 1$ take the values 0 and 1 in the sum. We assume that the correctness of different mappings is independent, i.e., $Pr(m_1, \ldots, m_k) = Pr(m_1) \cdot \ldots \cdot Pr(m_k)$. If no background knowledge is available, we apply for the apriori probability of a mapping to be correct the *maximum entropy principle* by setting $Pr(m_i = 1) = Pr(m_i = 0) = \frac{1}{2}$. We illustrate this calculation by an example.

Example 3.12 Consider the peer network in Figure 3.3 (a) with two mappings m_1 and m_2. Then

$$\begin{aligned} Pr(m_1 = 1 | f = 1) &= \frac{Pr(m_1 = 1)Pr(m_2 = 1)Pr(f = 1|m_1, m_2)}{Pr(f = 1)} \\ &+ \frac{Pr(m_1 = 1)Pr(m_2 = 0)Pr(f = 1|m_1, m_2)}{Pr(f = 1)} \\ &= \frac{1}{4Pr(f = 1)} \end{aligned}$$

For $m_1 = 1$, we have $Pr(f = 1|m_1, m_2) = 1$ when $m_2 = 1$ and $Pr(f = 1|m_1, m_2) = 0$ when $m_2 = 0$. Since $Pr(m_1 = 1) = P(m_2 = 1) = \frac{1}{2}$ and when assuming $\Delta = 0$ we have $Pr(f = 1) = \frac{1}{4}$, we therefore obtain $Pr(m_1 = 1 | f = 1) = 1$. Thus, it is not possible to obtain positive feedback when m_2 is not correct and we therefore may conclude that m_2 is correct.

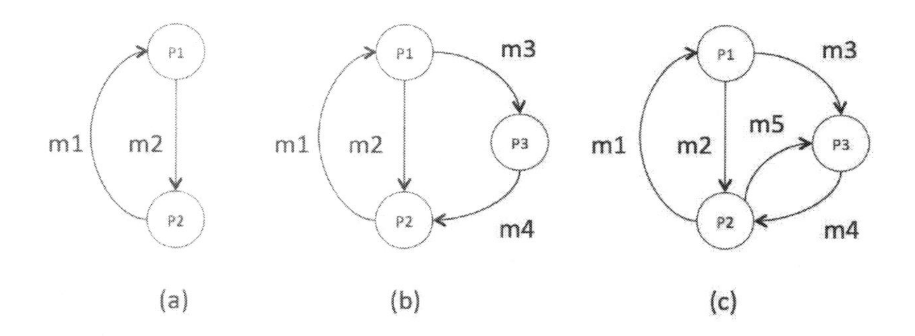

Figure 3.3: Three examples of peer-to-peer mapping networks with feedback.

The previous calculation is correct, provided the feedback is considered in isolation. In practice, we will have many feedbacks occurring at different peers, and the same mapping will influence the outcome of many different feedbacks. Thus, feedbacks and mappings are connected by complex dependencies, and the probabilistic inference for deriving mapping correctness from observed feedbacks becomes a complex task. In order to solve this task, the correlations between feedbacks and mappings are modeled graphically in a *factor graph*. Factor graphs are a generic technique for modeling complex, probabilistic correlation structures, where the joint probability distribution function decomposes into factors. Factor graphs support the efficient computation of marginal probabilities.

We illustrate the use of factor graphs with an example and refer the reader for a general formulation to Cudre-Mauroux et al. [2006].

Example 3.13 Assume a mapping contributes to the creation of two different feedbacks, as it is the case for mapping m_2 in Figure 3.3 (b). There, m_2 contributes to the cycle m_1, m_2, generating a feedback f_1 at peer p_1. At the same time, m_2 is equivalent to the mapping composed of m_3, m_4, generating a feedback f_2 at peer p_2. Therefore, the global probability distribution function for mapping correctness and feedback decomposes into two factors as follows:

$$Pr(m_1, m_2, m_3, m_4 | f_1, f_2) = Pr(m_1, m_2 | f_1) Pr(m_2, m_3, m_4 | f_2)$$

The computation of the marginal probability $Pr(m_2 | f_1, f_2)$ for determining the correctness of mapping m_2, given feedbacks f_1, f_2 can then be rewritten as follows.

$$\begin{aligned} Pr(m_2 | f_1, f_2) &= \sum_{m_1, m_3, m_4} Pr(m_1, m_2, m_3, m_4 | f_1, f_2) \\ &= \sum_{m_1, m_3, m_4} Pr(m_1, m_2 | f_1) Pr(m_2, m_3, m_4 | f_2) \\ &= \sum_{m_1} Pr(m_1, m_2 | f_1) \sum_{m_3, m_4} Pr(m_2, m_3, m_4 | f_2) \end{aligned}$$

This computation can be graphically represented as a factor graph as shown in Figure 3.4. For each feedback, we have a separate factor, represented by a square, and for each mapping, we have a variable indicating its correctness, represented by a circle. The mappings are connected to those factors to which they contribute in the feedback. The computation of marginal probability of a mapping to be correct, proceeds in two steps. We illustrate this for mapping m_2. First, all the mappings contributing to the feedback, with exception of the mapping m_2 for which we compute the marginal probability, send their probability of being correct as a message to the factor representing the feedback. These messages are of the form $\mu_{m_i \to f_j}$. The factor then computes the probability of mapping m_2 being correct, considering its feedback (analogous to the computation presented in Example 3.12). Next, it sends a message of the form $\mu_{f_j \to m_i}$ to m_2. m_2, finally, can compute its overall probability of being correct as the product of the individual probabilities determined for each factor. Since the computation of the marginal probability is based on sending of messages, this protocol is called a *message passing protocol*.

For acyclic factor graphs, it has been shown that this process produces, in a finite number of message exchanges, the exact marginal probabilities. However, in general, factor graphs derived from peer-to-peer mapping networks will not be acyclic. For that case, it has been shown that message passing in factor graphs still converges to a good approximation of the marginal probabilities and that these approximations are useful to assess the quality of mappings [Cudre-Mauroux et al., 2006]. Compared to other methods for solving factor graphs, like the junction tree algorithm, message

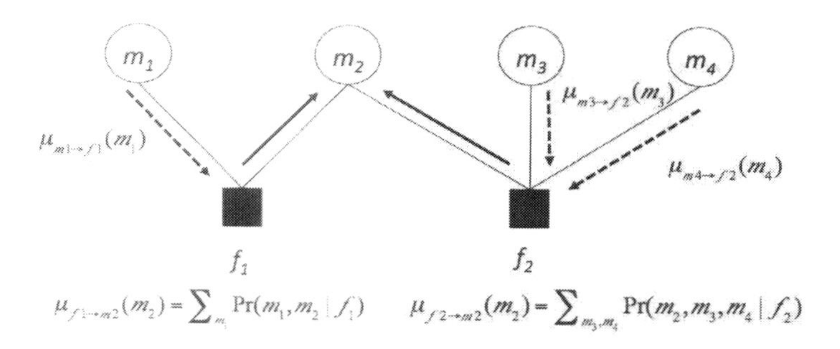

Figure 3.4: Example of a factor graph.

passing has the advantage that the computation can be easily distributed in a peer-to-peer network. Each peer computes the factors that belong to mappings that it is associated with, e.g., the peer being the source of the mapping, and the peers exchange the messages as dictated by the message passing algorithm.

Conclusion. Peer-to-peer data integration extends standard data integration by requiring a decentralized approach and, in particular, abandoning the concept of a global schema. Algorithms for peer-to-peer data integration have to take into account the decentralized nature of the system architecture and are inherently distributed. Since peers are autonomous, the discovery of unknown mappings and the consideration of the fact that mappings can be incomplete or incorrect plays a much more prominent role than in traditional data integration settings.

CHAPTER 4

Peer-to-peer Retrieval

Though peer-to-peer data management as introduced in the last two chapters is an important and relevant technical approach, most of today's data on the Web is accessed using the retrieval paradigm rather than the database querying paradigm. This is obviously the case for textual and media content, but also it is partially true for search on structured data. With the retrieval paradigm, searches produce ranked results of data or contents that match a query based on a similarity measure that models relevance. This is in contrast to the database querying paradigm where logical conditions are evaluated and results are sets of objects that satisfy those conditions. For very large collections of data and content, and in particular for exploratory search, the retrieval paradigm appears to be more appropriate.

Retrieval comprises always two aspects: first, the modeling of the retrieval task which comes down to modeling the informal concept of *relevance* through well-defined formal *similarity measures* and, second, the efficient implementation of similarity queries using these measures for large data collections. Whereas the first problem is at the heart of research in information retrieval, the second problem is closely linked to the problem of managing large databases, in particular, to the problem of indexing and implementing efficient search. Therefore, given our focus on peer-to-peer data management, we will introduce in the following techniques that have been devised for the second problem.

We will study two key technical problems that have been investigated to make searches on large collections of text documents efficient. First, we will consider the classical problem of full text retrieval, i.e., searches on the textual content of documents. Then, we will consider the problem of searches exploiting document structure since Web documents are indeed inherently structured, be it by the use of HTML or more generally XML. For both problems, we will consider of how, in particular, structured overlay networks can be employed to provide an efficient support for searches generated by retrieval queries.

4.1 FULL TEXT SEARCH

Full text search and retrieval are obvious candidates for using a peer-to-peer architecture, given the scale of textual data being made available on the Web today. While centralized Web search engines, such as Google or Bing, are widely accepted solutions for performing Web search, several considerations have motivated the investigation of alternative decentralized architectures for this task. These include avoidance of ranking manipulation or even censorship by central search engine providers, improved access to the Deep Web that is inaccessible to Web crawlers and the possibility to provide specialized and personalized search capabilities not supported by centralized search engines.

A key capability of a text retrieval system is full text search. For full text search, both documents and queries are represented as sets or bags of words from a given *term vocabulary* T. To generate this representation usually text preprocessing techniques such as stopword elimination and stemming are applied. Given a query $Q \subset T$, the basic operation of full text search, is to retrieve all documents $d \subset T$ for which $d \cap Q \neq \emptyset$. For efficiently implementing full text search the standard method in centralized retrieval engines is the use of *inverted files* [Zobel and Moffat, 2006].

Inverted files. An inverted file consists of a set of *inverted lists* of the form (t, L_t) where L_t is the *posting list* of term $t \in T$. The posting list is the sequence of *postings*. In the simplest case, a posting consists of a document identifier d; alternatively, it may consist of pairs $(d, f_{d,t})$, where $f_{d,t}$ is the frequency of *term occurrences* of term t in document d, called the *term frequency*. The postings are ordered in the posting list by the document identifier. In order to efficiently access the posting lists, an access structure for looking up terms t is provided. This access structure indexes tuples of the form (t, f_t, l_t), where f_t is the number of documents containing the term t, called the *document frequency* of t, and l_t is a pointer to the posting list of term t. In centralized systems, the access structure is implemented using standard indexing structures, such as B+-Trees.

Inverted files support efficient retrieval of all documents that contain a given term. This is the basic operation needed to evaluate retrieval queries for different retrieval models. For example, when the popular *tf-idf ranking* is applied, first, for all terms occurring in a query, the documents containing these terms are obtained. Then the contributions of all documents for each query term are aggregated, and a final similarity score between the query and the documents is computed.

There are good arguments why implementing Web-scale, full text search engine might not be feasible or at least only achievable when accepting compromises on search performance and quality [Li et al., 2003]. The fundamental limitations are the potential high bandwidth consumption and the difficulty to bring latency to acceptable levels. These reasons suggest that mainstream Web search might indeed not be feasible with a peer-to-peer architecture. Nevertheless, for specialized applications, a peer-to-peer architecture might be an interesting alternative. Furthermore, principles of peer-to-peer full text search might also be relevant for centralized, though highly distributed Web search engine implementations.

As for data management, also for full text search, a peer-to-peer architecture makes use of an existing overlay network. There exist two fundamentally different paradigms of how this can be realized.

tf-idf ranking. tf-idf ranking is the most popular method to evaluate text retrieval queries. Given a document d and a query $Q = \{t_1, \ldots, t_k\}$, tf-idf ranking computes a similarity value $sim_{tfidf}(d, Q)$ that is used to rank the search results. For computing the similarity, both the query and the document are represented as vectors $\vec{q}$ and $\vec{d}$ in the vector space over the term vocabulary T. Therefore, this is also called the *vector space model* of retrieval. These vectors are derived from the *term frequency* and the *inverse document frequency*. The term frequency $tf(t, d)$ counts how often a term t occurs in a document d. The inverse document frequency $idf(t)$ counts how often a term occurs in a document of the collection. Then

$$\begin{aligned} \vec{q}_t &= \log(1 + \frac{N}{idf(t)}) \text{ for } t \in T, 0 \text{ otherwise} \\ \vec{d}_t &= 1 + \log tf(t, d) \end{aligned}$$

N is the total number of documents. The similarity between a query and a document is computed as the *cosine similarity* of the two vectors:

$$sim_{tfidf}(d, Q) = \frac{\sum_{t \in T} \vec{q}_t \cdot \vec{d}_t}{\sqrt{\sum_{t \in T} \vec{d}_t^2} \sqrt{\sum_{t \in T} \vec{q}_t^2}}$$

There exist many variations of the weighting scheme used to generate the query and document vectors and to compute the tf-idf similarity measure. Their common underlying principles are that a document is more relevant to a query if a query term appears more frequently in the document, a term is less relevant for the query when it appears more often in the document collection, and a document is less relevant for a query when it contains more terms. tf-idf rankings can be efficiently computed using inverted files [Zobel and Moffat, 2006].

1. *Document partitioning:* With this approach, documents reside at the peers and searches are performed using flooding in an unstructured overlay network. At each peer, queries are evaluated against the local document collection, and results are returned to the query originator.

2. *Term partitioning:* With this approach, for each term a posting list of all term occurrences is produced. The term is then mapped to the identifier space of a structured overlay network, and the peer responsible for that identifier stores the posting list. A query is evaluated by retrieving the posting lists for all its terms using the lookup mechanism of the structured overlay network.

Naive implementations of both approaches are unlikely to scale up for very large document collections. Document partitioning will suffer from the high communication overhead for performing broadcast searches, and term partitioning will suffer from the problem of very long posting lists for frequently occurring terms.

We illustrate the problem for the case of term partitioning. Let a query Q consist of two terms t_a and t_b, and let L_a and L_b be the posting lists stored at two peers p_a and p_b. One possible processing strategy is to send the query first to p_a, p_a sends its posting list for t_a to peer p_b, and peer p_b computes the result, which it returns to the query originator. The processing is illustrated in Figure 4.1(a). The problem is that many posting lists will be very long as we show now.

We consider a Web size data collection with $s = 10^{10}$ documents, which corresponds to the number of documents indexed by the big search engine providers in the near future. Document identifiers are assumed to have 5 Bytes, which is largely sufficient to address such a document collection since $\log_2(s) \approx 33\ bits$. Using data from recent studies [Chierichetti et al., 2009], we assume that the average document length is $l = 500$ and that the size of the vocabulary is determined using Heap's law as $|T| = (s \cdot l)^\theta$ with $\theta = 0.7$. We do not consider stop word removal, as it is typical for web search engines. The same study also finds that the term frequency follows a Pareto distribution with $\alpha = 0.9$, i.e., the probability of the i-th term to occur is proportional to $i^{-\alpha}$. As a result, we obtain a term frequency of

$$f_{t_i} = \frac{1}{c_\alpha} \cdot i^{-\alpha}$$

where $c_\alpha = \sum_{i=1}^{|T|} i^{-\alpha} \approx 68$ with the parameters we assume. The most frequent term has then $f_{max} = 7.5 \cdot 10^{10}$ occurrences, which is consistent in the order of magnitude with numbers reported in Web search engines when searching for frequent terms.

For a term with a given frequency, we compute its document frequency in expectation, i.e., the number of documents it occurs in. The probability that a term does not occur once in a given document is $(1 - 1/s)$, and thus the probability that a term t_i, occurring f_{t_i} times, never occurs in a given document is $(1 - 1/s)^{f_{t_i}}$. Therefore, the probability of the term to occur in a given document is $1 - (1 - 1/s)^{f(t_i)}$, and the total number of documents the term occurs is

$$postings(t_i) = s \cdot \left(1 - (1 - \frac{1}{s})^{f_{t_i}}\right)$$

Using postings that consist of document identifiers only, the postings in a posting list are stored using *gap encoding*, i.e., the document identifiers are stored in sequential order and only differences among subsequent document identifiers are stored. No further compressions are considered. For a posting list of length l_p and a document identifier space of size l_{id}, the average number of bits needed to represent the differences is then $\log_2(l_{id}/l_p)$.

Example 4.1 Given identifiers of length 5 bytes and the posting list of the most frequent term with frequency f_{max} in our assumed document collection, on average, approximately 6.78 bits are needed for each posting. The following table lists the length of posting lists for terms t_i in megabytes.

i	*Length*
10^0	8471.
10^1	5666.
10^2	1370.
10^3	234.5
10^4	36.58
10^5	5.478
10^6	0.7991
10^7	0.1143
10^8	0.01613
10^9	0.002249
10^{10}	0.0003106

The result shows that for the most frequent terms the length of posting lists is in the range of gigabytes and even terms that are at frequency rank 10^7 still have posting lists of length in the hundreds of megabytes.

Considering that when processing a two term query in a distributed retrieval system in the naive way, as described above, the amount of data that needs to be transferred over the communication network corresponds to the size of the posting list of the less frequent among the two terms, distributed processing of retrieval queries is clearly problematic.

Therefore, many possible optimizations have been proposed to make Web scale peer-to-peer retrieval feasible. We discuss, in the following, approaches for term partitioning. The primary challenge for term partitioning is reducing the communication cost for processing multi-keyword queries.

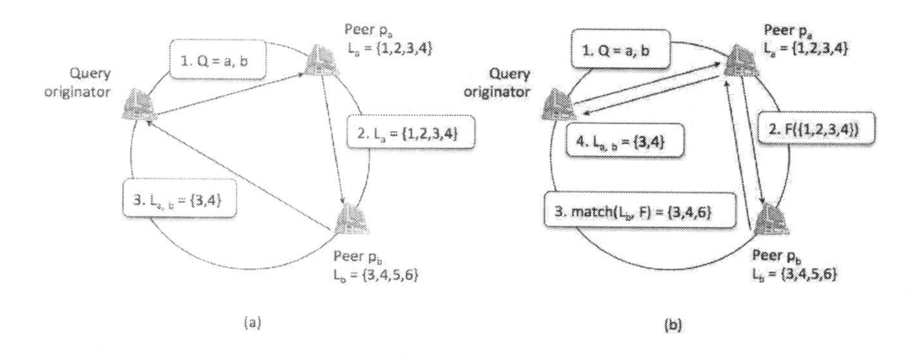

Figure 4.1: Illustration of processing a two term query using (a) the naive algorithm and (b) Bloom filters.

4.1.1 BLOOM FILTERS

One solution to reduce communication cost is based on the use of Bloom Filters [Bloom, 1970]. Bloom filters provide a more compact representation of the posting lists [Li et al., 2003, Reynolds and Vahdat, 2003]. They are a compact data structure for supporting set membership queries probabilistically.

Bloom Filters. A Bloom filter consists of a bit vector $\vec{b}$ of length m, initially set to 0. It uses k independent random hash functions $h_1, \ldots, h_k$ mapping elements from a domain T to integers in the range $[1, m]$. For creating the representation of a set $D \subseteq T$, for each element $t \in D$, the bits at positions $h_1(t), \ldots, h_k(t)$ of $\vec{b}$ are set to 1. For testing whether an element t' occurs in D, the bits at positions $h_1(t'), \ldots, h_k(t')$ are checked. If any of those bits is 0, t' can obviously not be contained in D. If all of them are 1, t' could still be a false positive. The probability of being a false positive is approximately equal to $(1 - e^{-k \cdot l/m})^k$, where $l = |D|$. To support removals of elements from D, a counter $c_i, i = 1, \ldots, k$, counting the number of times a bit has been set, can be maintained.

Query processing can exploit Bloom filters in the following way. Let the query Q again consist of two terms t_a and t_b. The query originator sends the query Q to p_a. The peer p_a computes a Bloom filter $\vec{b}(L_a)$ for its posting list L_a for term t_a and sends it to p_b. The peer p_b computes which elements of its own posting list L_b for term t_b match the Bloom Filter. It sends the result back to p_a, since the result may contain false positives. Finally, p_a computes the result and sends it to the query originator. The processing is illustrated in Figure 4.1(b). The technique can be extended to processing queries containing more than two terms [Reynolds and Vahdat, 2003].

We now investigate whether a reduction in communication cost can be achieved by using a Bloom Filter. Let us assume that the posting list L_a is the shorter one. Ignoring the transfer of the final result, which is the same for both algorithms, in the naive algorithm the remaining communication cost is the transfer of the posting list L_a. Using Bloom filters, the communication cost we have to compare to consists of the transfer of the Bloom filter $\vec{b}(L_a)$ and the back transfer of the result from p_b to p_a, including the false positives.

The false positive rate of a Bloom filter of size m, using k different hash functions, encoding a set of size l is approximately

$$fp(l, k, m) = (1 - e^{-\frac{k \cdot l}{m}})^k$$

So the total cost of using Bloom filters for query processing, apart from result transfer, is

$$cost_{BF}(bits, k, m) = m + fp(|L_a|, k, m) \cdot |L_b| \cdot bits = m + (1 - e^{-\frac{k|L_a|}{m}})^k \cdot |L_b| \cdot bits$$

where $bits$ is the average number of bits used when transferring back the result to p_a. We assume that the result set is small in comparison to the number of false positives, and thus estimate $bits$ as

$$bits(k, m) = \log_2(\frac{s}{fp(|L_a|, k, m) \cdot |L_b|}) ,$$

where s is the size of the document collection. In order to determine the optimal parameter settings m, k for using the Bloom filter, we have to minimize the expression

$$cost_{BF}(bits, k, m) = m + fp(|L_a|, k, m) \cdot L_b \cdot bits(k, m)$$

which can be done numerically.

Example 4.2 We show the results using document identifiers of length 128 bit in the posting lists. In the tables, the rows correspond to the size of the posting list L_a and the columns to the size of the posting list L_b. The entries are the ratio between the cost of the algorithm using Bloom filters and the naive technique.

	10^2	10^3	10^4	10^5	10^6	10^7	108
10^5	0.304688	0.244473	0.182492	0.121618			
10^6	0.267322	0.215157	0.170181	0.131335	0.0931637		
10^7	0.323387	0.262628	0.210399	0.166317	0.128389	0.0911962	
10^8	0.38881	0.317867	0.256976	0.205738	0.162622	0.125579	0.0893187

For 128 Bit identifiers, Bloom filters can result in considerable compression factors, as also found in previous works [Li et al., 2003, Reynolds and Vahdat, 2003]. For shorter identifiers, in particular identifiers with optimized length, e.g., 5 Bytes as considered earlier, the savings are however significantly lower. Processing using Bloom filters becomes even less efficient than the naive processing strategy.

4.1.2 SHORTENING OF POSTING LISTS

In order to avoid transfers of long posting lists, they can be split among multiple peers into *distributed posting list partitions* [Abiteboul et al., 2008]. The peer p_a, responsible for managing the postings of a term t_a, maintains in a *root block* an index of the partitions of a posting list. The root block consists of an ordered sequence of values $d_1, \ldots, d_k$ from the document identifier space, indicating where the posting list has been split. Initially, the list is empty and all postings are stored at peer p_a. When the length of the posting list exceeds a threshold, it is split into two sublists: the first, with document identifiers $d \leq d_1$, stored at peer p_a, and the second, with document identifiers $d > d_1$, stored at an overflow peer. The value of d_1 is inserted into the index stored at the root block. The overflow peers are adressed by a hash function that depends on the indexed term t_a and the index i of the partition. The same procedure is repeated whenever any of the distributed posting lists overflows. An

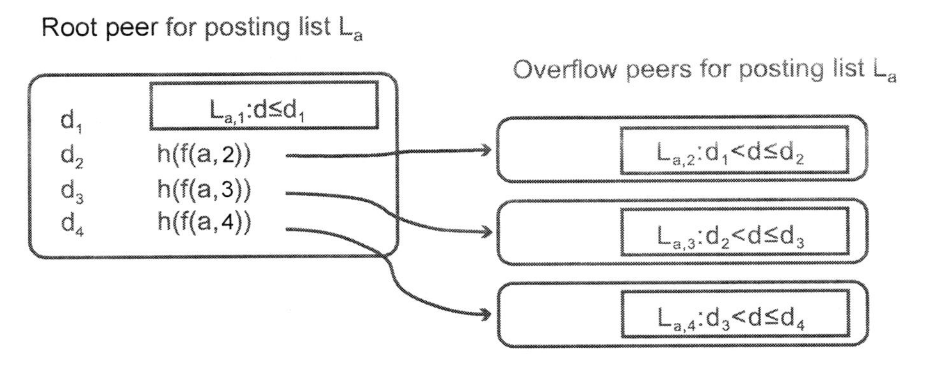

Figure 4.2: Splitting and distributing posting lists over multiple peers. The peers are addressed by a hashing value that is a function of the indexed term t_a and index i of the partial posting list $L_{a,i}$.

example of the resulting partitioning of the posting lists is shown in Figure 4.2. This approach bears resemblance with the B-tree index structure and can be generalized to a hierarchical data structure.

Having partitioned the posting lists over multiple peers opens the possibility to process queries in parallel, which reduces query latency. We describe the processing of a two term query. First, the query originator retrieves the two root blocks, corresponding to the query terms. From their indices, it can determine all pairs of partitions that are overlapping, i.e., have potentially postings in common. The different overlapping pairs of posting list partitions can then be processed independently in parallel, either at the query originator or by the peers holding the posting list partitions.

Another strategy to reduce the cost of processing multiple keyword queries is based on indexing combinations of multiple terms [Skobeltsyn et al., 2007]. The basic idea is to trade-off the increased cost for constructing a much larger set of posting lists for reduced communication cost during query processing. Since the number of term combinations grows very quickly with the size of the term combination, only combinations that are potentially useful for processing queries are selected for indexing. In this approach, different types of term combinations are considered as keys for indexing.

- *Non-discriminative keys:* These are terms or term combinations that appear in more than a maximal number s_{max} of documents.
- *Discriminative keys:* These are terms or term combinations that appear in less than a maximal number s_{max} of documents.
- *Intrinsically discriminative keys:* These are terms or term combinations that appear in less than a maximal number s_{max} of documents and do not contain any proper subset of terms that is a discriminative key.

For constructing the index, only non-discriminative and intrinsically discriminative keys are used. The length of posting lists is limited to s_{max} documents for non-discriminative keys, and only the documents having the highest ranking are included into the posting list. In addition, other criteria such as term proximity can be used to exclude further term combinations from indexing. The set of terms used for indexing is called *highly discriminative keys*.

Experiments show that the overall index size is by an order of magnitude larger than for single term indexing, but that the overall communications cost is also reduced by orders of magnitudes, due to reduced query traffic. This happens without loosing much query precision as a result of limiting the size of posting lists for non-discriminative keys.

4.1.3 CACHING

In order to further reduce indexing and query processing cost, various *caching* techniques have been considered. In the simplest case, posting lists or Bloom filters that are transferred to another peer are cached at that peer for later reuse as long as storage capacity is available [Reynolds and Vahdat, 2003]. Caching can also be used to selectively construct posting lists for multi-term keys, driven by the usage frequency of those keys in queries [Skobeltsyn et al., 2007]. This avoids the problem of indexing many unnecessary term combinations when using the basic multi-term indexing approach described before. In this approach, multi-term keys can be in three different states:

1. *Active:* A posting list is maintained for the key.

2. *Candidate:* The key is intrinsically discriminative and has appeared in a query. A query statistics is maintained for the key.

3. *Inactive:* The key is not intrinsically discriminative or has never occurred in a query.

When a query is processed, the peer searches for the longest term combinations that are active keys, i.e., for which a posting list is available. During query processing, also, the query statistics is updated, and keys might get promoted from inactive to candidate status, or from candidate status to active status. The detailed algorithm is shown in Algorithm 6.

Algorithm 6 proceeds by extracting the subkeys of decreasing length from the query Q, starting from a maximal key length l_{max} (line 3 and 4). It first checks whether the selected key is not a proper subkey of an active key (line 5); in which case, nothing needs to be done. For those keys the responsible peer is contacted (line 7) to obtain the status of the key and to update the access frequency, in case the key is a candidate key (line 4 of Algorithm 7).

If the key is active, the posting list of the key is added to the result set for further processing at the query originator (line 9). If the key is inactive, it is considered as a candidate for later activation (line 12). For inactive keys or active keys that are discriminative, it is checked whether they are subkeys of keys that are considered as candidates for activation (lines 14 and 15). In both cases, the superkeys do no longer qualify for activation (line 16). Once the complete set of subkeys of the query keys is explored, the keys that remained as candidates for activation are moved to the candidate status (line 24).

Algorithm 6: search(Q)

```
1  K = ∅; C = ∅ ;                         /* K active keys, C new candidate keys */
2  R = ∅; ;                                                      /* R result */
3  for i = min(l_max, |Q|) downto 1 do              /* l_max maximal key length */
4  |  for k ⊂ Q with |k| = i do
5  |  |  if k ⊄ k', k' ∈ K then           /* k not proper subkey of active key */
6  |  |  |  p = get(h(k));                     /* find peer responsible for k */
7  |  |  |  state = p.probe(k);
8  |  |  |  if state = active then
9  |  |  |  |  K = K ∪ {k};
10 |  |  |  |  R = R ∪ {p.L(k)};                 /* L(k) posting list for key k */
11 |  |  |  end
12 |  |  |  if state = inactive then                 /* k potential candidate key */
13 |  |  |  |  C = C ∪ {k};
14 |  |  |  end
15 |  |  |  if state = inactive or (state = active and |p.L(k)| ≤ s_max) then
16 |  |  |  |  while exists k' ∈ C such that k ⊆ k' do
17 |  |  |  |  |  C = C \ {k'}
18 |  |  |  |  end
19 |  |  |  end
20 |  |  end
21 |  end
22 end
23 for k ∈ C do
24 |  p = get(h(k));                            /* find peer responsible for k */
25 |  p.createCandidate(k);                      /* move key to candidate status */
26 end
```

4.1.4 FAGIN'S ALGORITHM

In information retrieval, usually, only some top ranked documents need to be returned as result. This observation is the basis for a different type of optimization of text search in peer-to-peer retrieval, based on top-k retrieval methods such as Fagin's algorithm [Suel et al., 2003].

In its basic form, the algorithm is not suitable for a peer-to-peer environment. It would require a large number of message exchanges among the peers storing the posting lists. Therefore, the algorithm needs to be modified in order to minimize communication cost [Suel et al., 2003]. We describe this modification for the case of a two-term query with query terms t_a and t_b.

Algorithm 7: probe(k)

```
1  if inactive(k) then
2  |  return inactive
3  end
4  freq(k) = freq(k) + 1;
5  if candidate(k) then
6  |  if freq(k) > f_min then
7  |  |  create posting list for k;
8  |  |  return active
9  |  else
10 |  |  return candidate
11 |  end
12 else
13 |  return active
14 end
```

Fagins Algorithm. Fagin's algorithm is an efficient method to retrieve the highest ranked elements in a document collection when the ranking function is composed of multiple features [Fagin, 1999]. It is based on the intuition that a document with a high total rank is also highly ranked for at least one feature. In order to apply the algorithm, the documents have to be first sorted according to the feature values. The algorithm then proceeds in two phases. The sorted document list are traversed from the top in a round-robin fashion. This is continued till k documents are found that occur in all lists. For those documents that have, at that point, not been found in all lists, in a second phase, the documents are looked up in the other lists, using random accesses. Finally, the scores of all documents found are computed, and the k highest ranked documents are returned as result. The algorithm is guaranteed to find the top-k elements. It is illustrated in Figure 4.3 for the case of two features.

Peer p_a that is storing the shorter posting list sends the first β elements of its posting list and the minimal score f_a of these elements to peer p_b. β is a design parameter. Peer p_b looks up the elements received from p_a in its own posting list and computes the combined scores, which is the sum of the individual scores. It retains the k elements with the highest scores and determines the minimal score f_b among these. Next, p_b selects from its β first elements, which have not yet been combined with an element from p_a, those that have a score larger than $f_b - f_a$. It sends them to p_a, together with the scores of the top-k elements it has found so far. Note that an element with score less than $f_b - f_a$ can no more become top-k element as it would have to be combined with an element from

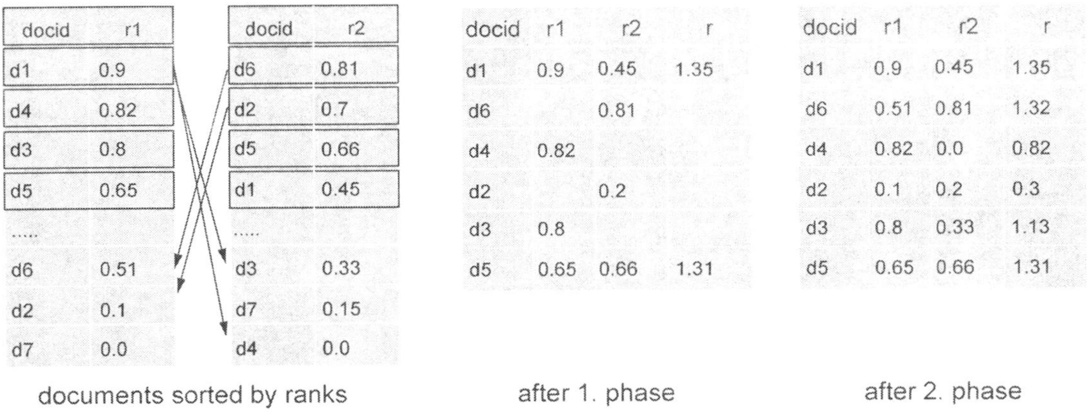

docid	r1
d1	0.9
d4	0.82
d3	0.8
d5	0.65
.....	
d6	0.51
d2	0.1
d7	0.0

docid	r2
d6	0.81
d2	0.7
d5	0.66
d1	0.45
.....	
d3	0.33
d7	0.15
d4	0.0

documents sorted by ranks

docid	r1	r2	r
d1	0.9	0.45	1.35
d6		0.81	
d4	0.82		
d2		0.2	
d3	0.8		
d5	0.65	0.66	1.31

after 1. phase

docid	r1	r2	r
d1	0.9	0.45	1.35
d6	0.51	0.81	1.32
d4	0.82	0.0	0.82
d2	0.1	0.2	0.3
d3	0.8	0.33	1.13
d5	0.65	0.66	1.31

after 2. phase

Figure 4.3: This figure illustrates Fagin's algorithm for finding the top-2 documents when two features are used to rank the documents. In a first phase, the two sorted lists are traversed till two documents are detected in both lists, namely d_1 and d_5. In a second phase, the other documents are looked up using random accesses, indicated by arrows. Finally, the combined scores are computed, and the two elements with highest scores, namely d_1 and d_6, are returned as result.

p_a that has a score less than f_a. Thus, the combined score would be less than f_b, which is smaller than the minimal score of the current top-k elements. Finally, p_a selects the overall top-k elements from the data received from p_b. The block size β can be determined from historical data or through sampling. If it is too small to obtain k elements, a second round of exchanges can be performed.

4.1.5 PEER SUMMARIES

Another strategy to reduce communication cost in peer-to-peer text retrieval is to adopt a two step approach in indexing and search [Bender et al., 2005]. In a first step, peers publish and search summaries of their local document collections. In a second step, full text searches are processed only for selected peers that have promising content, by using their local indices. This approach relies on the capability of providing good collection summaries for peer document collections.

For summarizing a collection, similar measures such as tf-idf ranking at document level are used at the peer collection level [Callan et al., 1995]. Given a peer p_i and a term t, the following two measures are used to determine the relevance of a term with respect to the document collection of p_i and the distinctive power of the term. The first measure is the *collection document frequency* of the term, and it corresponds to the term frequency in the standard tf-idf ranking:

$$cdf_{i,t} = df_\alpha + (1 - df_\alpha) \cdot \frac{df_{i,t}}{df_{i,max} + 50 + 150 \cdot \frac{|T_i|}{|T_{avg}|}} .$$

$df_{i,t}$ is the document frequency of term t in the collection of p_i, i.e., the number of documents in which t occurs at p_i, and $df_{i,max}$ the maximal document frequency at peer p_i. $|T_i|$ is the number of distinct terms at peer p_i, and $|T_{avg}|$ is the average number of distinct terms found at all peers. df_α is the default frequency and set to $df_\alpha = 0.4$.

The second measure is the *inverse collection frequency* of the term and it corresponds to the inverse document frequency in the standard tf-idf ranking:

$$icf_t = \frac{\log(n + 0.5)}{n_t \cdot \log(n + 1)} .$$

n is the total number of peers and n_t the number of peers that contain t.

For answering a query $Q = \{t_1, \ldots, t_k\}$, the similarity to a collection at p_i is then computed as

$$sim(Q, i) = \frac{1}{n} \sum_{t \in Q} sim_\beta + (1 - sim_\beta) \cdot cdf_{i,t} \cdot icf_t .$$

sim_β is the default similarity and set to $sim_\beta = 0.4$.

For obtaining the various quantities needed for evaluating this similarity measures, a structured overlay network is used. The overlay network indexes terms t. Peers responsible for a term t maintain a posting list of the occurrences of the term at peers p_i, together with their collection document frequencies $cdf_{i,t}$. The length of the posting list can be used to determine n_t. Peers can locally compute $cdf_{i,t}$, provided they know $|T_{avg}|$ and publish this information to the overlay network. $|T_{avg}|$ as well as n can be determined, using methods for aggregation query processing introduced earlier in Section 2.2.3.

Conclusions. Given the interest into the problem of full text retrieval for large scale document collections in a peer-to-peer setting, a variety of techniques have been studied to make, in particular, the approach of term partitioning feasible. One aspect that we did not cover concerns the process of indexing, i.e., efficiently constructing inverted files distributed in a peer-to-peer network. To that end, different techniques such as aggregation of messages into larger blocks and congestion control have been considered [Klemm et al., 2006].

Approaches based on document partitioning are typically used in conjunction with unstructured overlay networks. We will provide an introduction to such techniques in the forthcoming chapter on semantic overlay networks.

4.2 STRUCTURED DOCUMENT SEARCH

Due to the wide-spread use of XML as standard document format on the Web, increasing amounts of XML data are expected to be shared on the Internet. This has also raised the interest of managing large collections of XML documents in peer-to-peer systems. The support for the document structure in XML documents introduces novel requirements on indexing and searching in peer-to-peer networks, in addition to those imposed by supporting search on the text content.

Structured querying of XML documents is supported through standardized query languages, notably XPath and XQuery. XPath supports the concept of search on *paths* in a document tree using regular expressions. This is a capability that requires query processing capabilities that are significantly different from those used for standard text retrieval and relational query processing. XQuery builds on XPath. It introduces well-known concepts from SQL, in particular, the possibility perform joins, as well as the possibility to search for complex tree patterns.

Supporting these extended query functions already poses significant technical challenges in centralized settings, which are excarberated in peer-to-peer settings. We will discuss, in the following, some key approaches by focusing on those aspects that are peculiar to decentralized indexing and querying. We refer the reader to the literature for details of underlying techniques, borrowed from centralized query processing.

4.2.1 PATH INDEXING

Processing of path queries in documents can be understood as an extension of full text retrieval on text as well as a special form query processing on structured data. As a consequence, techniques from full-text retrieval, e.g., for regular expression search, and structured data management, e.g., for processing path queries in hierarchical or object-oriented data models, are applicable [Luk et al., 2002]. We now investigate of how such techniques can be adopted to the setting of peer-to-peer data management and, in particular, of how structured overlay network can be exploited for that purpose.

We introduce some basic notations and assumptions for the exposition of the following. We will denote paths that occur in XML documents using simplified *XPath expressions*. In particular, we will treat the terms occurring in the text of a content at the leaf level the same as XML element names. We will call these XPath expressions *path patterns*. When a path pattern matches a document, we will call this a *path occurrence*. For example, for the document given in Figure 4.4, some path occurrences of path patterns in the document are

1. Complete paths from the root to a leaf: `/book/author/last/Aberer`
2. Subpaths in the document tree: `editor/first/Tamer`
3. Paths containing *descendant operators* or *wildcards*: `book//last`, `book/*/last`
4. Single elements or terms: `editor`, `Tamer`

For indexing XML documents using structured overlay networks, the main design choice is on the mapping of path patterns to identifiers from I. This enables efficient lookup of documents in which corresponding paths occur. Since the same path can occur in many different documents, a peer will maintain, in general, a *posting list* for each indexed path pattern, containing all occurrences as the postings. This essentially corresponds to constructing a distributed inverted file for text indexing, as discussed in the previous section.

```
1  <book>
2      <author>
3          <first>Karl</first>
6          <last>Aberer</last>
9      </author>
10     <editor>
11         <first>Tamer</first>
14         <last>Ozsu</last>
17     </editor>
18     <title>
19         Peer-to-peer_Data_Management
20     </title>
21     <year>2011</year>
24 </book>
```

```
<book>
    <author>
        <first>Serge</first>
        <last>Abiteboul</last>
    </author>
    <editor>
        <first>Tamer</first>
        <last>Ozsu</last>
    </editor>
    <title>
        XML_Data_Management
    </title>
    <year>2012</year>
</book>
```

Figure 4.4: Sample XML documents d_1 and d_2; the numbers indicate the position of the element tag and contents when the document is sequentially parsed.

Indexing single elements and terms. In the simplest case, only the element names and content terms are used to generate identifiers by applying a hash function [Abiteboul et al., 2008, Galanis et al., 2003]. For each occurrence of such a path pattern e, where e is an element name or term, a posting $(e, (p, d))$ is generated, where d is the document identifier and p the identifier of the peer p that stores the document. The posting may include, in addition, structural information. *Positional information* indicates the position of the element in the document and the level at which it appears in the XML hierarchy, resulting in multiple postings for different occurrences of the same element or term [Abiteboul et al., 2008]. *Structural summaries* consist of a list of all paths leading to the element or term, resulting in a single posting for multiple occurrences of the same element or term [Galanis et al., 2003].

Example 4.3 Consider the sample document d_1 given in Figure 4.4 and the path pattern `first`. Using positional information, we would obtain the postings (`first`, $p, d_1, (3, 5, 3)$) and (`first`, $p, d_1, (11, 13, 3)$) where the positional information (s, e, h) contains the position of the start and end of the element s and e and the level h in the tree hierarchy. Using structural summaries would result in a single posting (`first`, p, d_1, {`/book/author`, `/book/editor`}).

For processing path or tree queries, all the elements names and terms occurring in the query are looked up, and the posting lists are returned to the query originator [Abiteboul et al., 2008]. The query originator performs post-processing to filter out the documents that match the query. This processing step can use standard techniques for optimized processing of tree queries, such as holistic twig join [Bruno et al., 2002]. Optimizations for handling very long posting lists are analogous to the techniques used for efficiently handling inverted files in text indexing introduced in Section 4.1. Also optimizations used for multi-attribute query processing as introduced in Section 2.2.1 can be applied in this context.

For processing simple path queries of the form `/a1/a2 ... /ak` a different strategy can be pursued [Galanis et al., 2003]. Only the last element ak in the path expression is looked up. Using the structural summaries, the occurrences that can potentially match the query path are selected from the posting list and returned to the query originator for post-processing. This approach can be extended to queries of the form `/a1[q1]/a2[q2] ... /ak[qk]` where at each level filter conditions `q1, q2, ..., qk` can be added. These filter conditions are path queries themselves. When processing such a query, also, the last elements in the paths `q1, q2, ..., qk` are looked up. In this case, optimizations used for multi-attribute query processing can again be applied.

Indexing path patterns. In order to make query processing more efficient, more complex path patterns can be indexed. This may help to avoid retrieval of multiple posting lists during query processing and, at the same time, reduce the length of the posting lists, similarly to multi-term indexing for full text retrieval. This comes at the price of generating more indexed keys and the total size of the distributed index.

A simple strategy is indexing all subpaths occurring in a document [Skobeltsyn et al., 2005]. This approach would for the sample document d_1 and the path `book/year/2011` generate postings for the path patterns `book/year/2011, year/2011, 2011, book/year, year` and `book`. This incurs an overhead of $O(l)$ compared to indexing only single element names and terms, where l is the average document height in the document collection. The path patterns are mapped to an identifier space $\{0, 1\}*$ using a *prefix-preserving hash function*. Such a hash function maps a path `q1` that is a prefix of another path `q2` to an identifier h(`q1`) that is a prefix of the identifier h(`q2`). An identifier resulting from hashing a path pattern may be prefix of different peer identifiers. The path pattern can then be stored at any of these peers.

Example 4.4 Assume that element name in our example documents are encoded by 3 bit identifiers and paths are encoded by concatenation. For example, this could result in the following mappings: h(`book`) $= 000$, h(`book/year`) $= 000101$. If there exist different peers with identifiers that share a prefix with a hashed path pattern, e.g., $id_1 = 0000$ and $id_2 = 0001$, then such a pattern can be stored at any of these peers, as it is the case for h(`book`). h(`book/year`) can, however, only be stored at the peer with identifier id_2. For finding the postings corresponding to h(`book`), then all peers that have a identifier prefix 000 need to searched.

In general, for retrieving all postings corresponding to a specific path pattern, `q = a1/a2 ... /ak` a range search for h(`q`) is performed using algorithms described earlier in Section 2.1.5. This corresponds to evaluating a path query of the form `q/*`. For processing a general XPath query, the longest contiguous path `q = a1/a2 ... /ak` of the query is selected, i.e., the longest subpath in the query that contains no descendant or wildcard operators. Then a range search for h(`q`) is performed. Further optimizations, in order to reduce the number of range searches, are possible using caching [Skobeltsyn et al., 2005].

Exploiting Document Type Definitions. When a Document Type Definition (DTD) is known, indexing of path patterns can be performed more selectively [Garcés-Erice et al., 2004]. We assume that a fixed DTD is given, and from query logs, a set of frequently used XPath query templates Q can be derived. The query templates in Q have the form `q($1, ..., $k)`, where `$1, ...,` `$k` are variables that are bound to the values in the text contents of the document. For example, for document d_1 in Figure 4.4 the query template `/book/author[first/$1][last/$2]` binds the variables `$1=Karl` and `$2=Aberer`. Using this binding, a XPath query can be constructed that matches exactly the values found in the document, e.g., for the previous example, the query `/book/author[first/Karl][last/Aberer]` would be constructed. This query is then hashed to generate an identifier for inserting all documents, satisfying the query into a structured overlay network.

The set of query templates Q includes also a most specific query template `qd` that reflects the complete structure and content of documents corresponding to the DTD.

Example 4.5 For the documents d_1 and d_2 the query template
`/book/editor[first/$1][last/$2]`
results in the same identifier h(`/book/editor[first/Tamer][last/Ozsu]`) for the two documents. The most specific query template for these documents, considering the naturally implied DTD, would be

```
qd = /book[author[first/$1][last/$2]]
          [editor[first/$3][last/$4]]
          [title/$5]
          [year/$6]
```

The identifiers generated for this query template after binding the variables uniquely identify the documents.

When multiple query templates are used, by exploiting the query subsumption relationship, a query hierarchy is constructed. A path query `q1` subsumes a path query `q2`, written as `q1` $\sqsubseteq$ `q2` if for all documents d, if `q1` matches d, then also `q2` matches d. Thus, the most specific query template `qd` subsumes all other queries. From the subsumption relationship, a query lattice is constructed, where the most specific query is the root, as it subsumes all the others.

Identifiers for insertion into the structured overlay network are generated for the query templates in the query lattice that match a document. The posting for the identifier, generated for the most specific query template, points to the document. All other postings point to the postings generated from the more specific query template contained in the query lattice.

Example 4.6 We consider the following set of query templates.

```
q1 = /book[author[first/$1][last/$2]]
q2 = /book[editor[first/$1][last/$2]]
q3 = /book[editor[last/$4]]
q4 = /book[title/$5][year/$6]
q5 = /book[title/$5]
q6 = /book[year/$5]
```

Then the subsumption relationships are qd $\sqsubseteq$ q1,q2,q4, q2 $\sqsubseteq$ q3, and q4 $\sqsubseteq$ q5,q6. The template q3 generates for documents d_1 and d_2 the identifier h(/book[editor[last/Ozsu]]) and the template q2 generates the identifier h(/book[editor[first/Tamer][last/Ozsu]]), for both d_1 and d_2. The node responsible for h(/book[editor[first/Tamer][last/Ozsu]]) would store the following posting list

$$(h(\texttt{/book[editor[first/Tamer][last/Ozsu]]}), \{h(\texttt{qd1}), h(\texttt{qd2})\}) .$$

where qd1, qd2 are the identifiers of the most specific query templates for documents d_1 and d_2.

The approach supports processing of queries, where all text values are bound. This corresponds to the case of processing multi-attribute queries with exact match in Section 2.2.1. For processing a query, first, the corresponding query template is located in order to determine the starting point for processing. If it does not exist, the most specific query in the hierarchy that the user query subsumes is identified. Then the query is iteratively resolved by following the postings, leading to the root and thus the document.

Example 4.7 If the query is

/book[editor[last/Ozsu]][year/2011]

then it subsumes query templates q3 and q6. So documents can be searched by starting from the peers holding posting lists related to these query templates.

Many other approaches of structurally indexing XML documents have been proposed. Large XML documents can be distributed over multiple peers [Bonifati and Cuzzocrea, 2006]. Document fragments are indexed in the structured overlay network by using path expressions. Retrieval-oriented approaches introduce the notion of *path similarity,* using paths as a feature to evaluate document similarity [Winter and Drobnik, 2009].

4.2.2 FILTERING

In order to improve the efficiency of XML query processing, the document structures occurring in large XML document collections can be summarized using *filters*. This is similar to the idea of using Bloom filters for full text retrieval. Filters are compact representations of document collections for efficiently testing whether a query potentially matches any of the documents in the collection. More formally, given a set of documents D, a filter $F(D)$ satisfies the following property: for any query Q, if the query does not match the filter $F(D)$, then the query does also not match any of the documents in D. In other words, testing a query against a filter may produce false positives, but never false negatives.

In peer-to-peer systems, filters can be used to reduce communication cost during search. The principle works as follows. Suppose a peer p_1 searches for documents matching a query Q, and another peer p_2 potentially stores matching documents. Instead of immediately sending a search request to peer p_2, it can first test whether sending a message to p_2 is worthwhile, using a filter that it earlier received from p_2. Assuming that filters are comparably small, the effort of distributing filter information in a peer-to-peer network can be kept small.

Another important property of filters is their composability. Given two document collections D_1 and D_2, then there exists an operation op such that

$$F(D_1 \cup D_2) = op(F(D_1), F(D_2)) .$$

Using this property, filters can be organized in hierarchical data structures. This property can be exploited in order to optimize query routing in overlay networks. Filters can help in query processing in all types of overlay networks.

- In an unstructured overlay network, each peer can maintain for each outgoing link a filter for all documents that are reachable with a given number of hops, the *horizon* [Koloniari et al., 2004]. Filters can be maintained for all horizons up to the time-to-live that is used in message routing. When a query is routed, for each outgoing edge, the filter with an horizon corresponding to the remaining time-to-live is used to test whether along that edge any documents potentially matching the query can be found.

- In a superpeer network, a set of superpeers serve as roots for filter trees that aggregate subsets of the peer's document collections hierarchically [Koloniari et al., 2004]. A query that arrives at any superpeer is first flooded among all superpeers and, for those where the filter at the root matches the query, is forwarded down the filter tree to peers that contain potential query matches.

- In a structured overlay networks, filter trees can be maintained as document summaries for all documents containing a specific element. This is similar to the approach of using path summaries [Galanis et al., 2003]. For scalability, the nodes of the filter tree can be distributed in the overlay network, by hashing identifiers of tree nodes to the identifiers space of the overlay network.

Bloom filters are one possible type of filter that can be used to support XML query processing [Koloniari and Pitoura, 2004]. Given an XML document d, we consider the sets of elements E_j, occurring at the same level j. The level of root elements is 1. For each of those sets E_j, a Bloom filter B_j is generated. Thus, a document with depth l is represented by l Bloom filters $B_1, \ldots, B_l$.

In addition, a filter B_0 for all elements occurring in the document is constructed. Filters representing different documents can be merged using a bitwise OR operation on the Bloom filters at the corresponding levels. In that way also, document sets can be represented.

For checking whether a path query matches the document, first, it is checked whether all elements in the query occur in the document by using the filter B_0. For paths starting from the root, next, it is checked whether all elements occur at the required levels. For subpaths `e1/e2/ .../ek` of queries, where the level information for elements is not known , e.g., following an // operator, a level s is searched such that each element `ei` matches the filter B_{i+s-1} for $i = 1 \dots k$.

Another technique for constructing filters using Bloom filters is based on an encoding of document paths [Koloniari and Pitoura, 2004]. Also, other types of filters have been used, such as *dataguides* to summarize the common structure of document collections [Rao and Moon, 2009, Sartiani et al., 2005], or the use of *irreducible polynomials* to characterize XML document structure [Rao and Moon, 2009]. In the latter approach, filter operations are based on algebraic manipulations. Filter composition is based on constructing the least common multiple of the polynomials, whereas matching of queries requires testing of divisibility.

Filters can also be used to address the problem of handling long posting lists in XML indexing [Abiteboul et al., 2008]. Given two postings lists, $L_{\mathtt{a}}$ and $L_{\mathtt{b}}$, for elements a and b, the peer storing $L_{\mathtt{a}}$ creates a filter $F(\mathtt{a})$ that the peer storing L_b can use to quickly test which of its elements b are potentially contained in an element a. In other words, it can quickly test for a subpath `a//b` occurring in a query, without requiring the transmission of the complete posting list $L_{\mathtt{a}}$. The construction of the filter $F(\mathtt{a})$ is based on Bloom Filters.

Conclusion. Indexing XML documents in peer-to-peer networks applies a blend of techniques both inspired from full text retrieval, such as inverted files and Bloom filters, as well as from structured data management, such as multi-attribute query processing and the use of schema information. This reflects the nature of XML documents which contain both textual content and structural information, but also the dual nature of paths in XML documents, which can be both interpreted as strings, and thus are amenable to text processing techniques, as well as hierarchical data structures, and thus are amenable to structured data management techniques.

CHAPTER 5

Semantic Overlay Networks

Semantic overlay networks (SON) extend on the idea of clustering related data in a structured overlay network for efficient retrieval, which we extensively studied in Section 2.1 for range partitioning. We adopt in this chapter a simple understanding of the notion of semantics. We assume that the semantics of information objects, both structured data or unstructured content, is given by a model that allows to express semantic proximity, i.e., whether two objects have a similar meaning. This model is given by mapping of the information objects into a metric space, in which the distance function captures the semantic proximity. Resources and interests of peers can then be equally modeled as points or regions in such a semantic space. The overall goal of constructing semantic overlay networks is to extend or modify a peer-to-peer overlay network structure, such that semantic locality is achieved. As a result, peers with similar resources or interests are better connected. This has two effects. Peers can find more likely content relevant to their needs in their immediate neighborhood. And when searching for a specific type of content, the results tend to be clustered in a region and can thus be more efficiently retrieved.

Figure 5.1 demonstrates the main principles of how semantic overlay networks work. The rectangle symbolizes a semantic feature space, in that case of two dimensions. Peers with similar resources and interests are clustered together. They are locally connected by short-range links. When a peer p is searching for a concept c, it greedily routes in the semantic space towards the target zone using its long-range links, e.g., passing through a peer p_r. Peers sharing the same interests are clustered together. It suffices that one or a few peers in such a cluster are connected by a long-range link towards their region of interest. In the figure, such a link would be maintained by peer p_c. When a new peer p_n arrives, it might physically not be connected to peers that share its interests but, as shown in the figure, to a remote peer p. However, it can search for its interest profile using the semantic overlay network and locate and connect to peers with common interests, as indicated by the broken arrow showing a new link.

The literature has produced a very rich and diverse set of ideas for constructing semantic overlay networks. A main reason for this diversity is the large number of design dimensions the basic concept explained above leaves for concrete realization. In the following, we will identify the key dimensions in order to better characterize different approaches that will be introduced in more detail later.

5.1 CONCEPT SPACES

We find different types of features that are used to characterize the semantics of a resource.

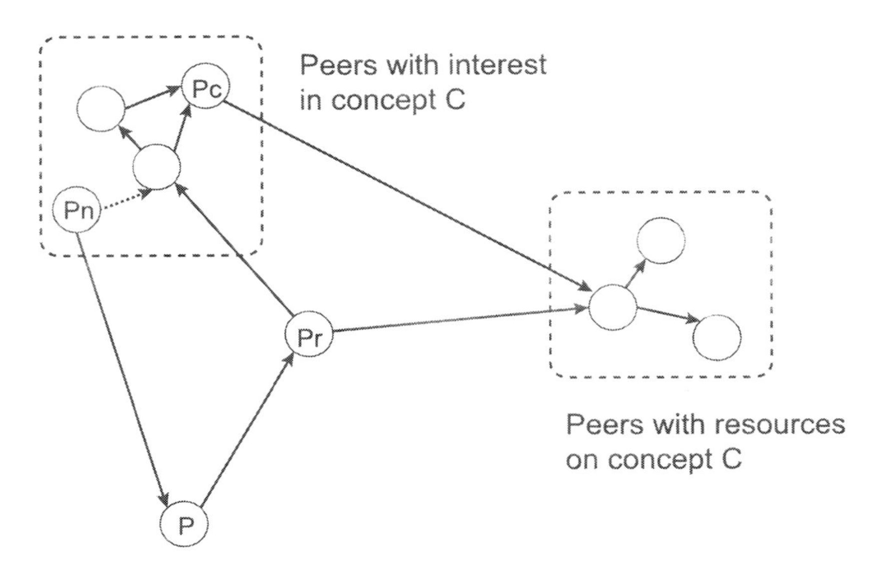

Figure 5.1: Illustration of semantic overlay networks.

1. *Categorization.* Resources are assigned with one or more predefined categories. The categories can be application-specific or be taken from some general-purpose classification schemes or ontologies.

2. *Full text description.* Resources are are annotated using a full text description. The semantics of such a full text description is usually derived from the statistical features of the text such as the tf-idf measure used in full text retrieval.

3. *Multimedia features.* If the resources consist of media files, such as image or audio, feature vectors can be extracted using content analysis tools.

It is important to note that the boundary between these types of concept spaces might be sometimes fuzzy. For example, there exist ontologies that are rooted in natural language, such as Wordnet [Miller, 1995], and are used for categorization. At the same time, some recent forms of textual annotation, such as tagging [Christiaens, 2006], bear more the character of a categorization approach than of a full text description.

The concept space may exhibit some additional internal structure. The two most common cases found for the construction of semantic overlay networks are

1. *Flat concept spaces.* They exhibit no internal structure.

2. *Hierarchical concept spaces.* Organize the concepts in an hierarchical structure, either a tree or a tree with some additional relationships such as terminological relationships.

More general internal structures such as lattice structures or general graph structures can be envisaged but have so far not been adopted in the construction of semantic overlay networks.

Finally, and most importantly, the concept space is equipped with a *similarity function* that measures the semantic similarity of two concepts and is at the heart of creating an overlay network structure with strong locality properties. We find the following two basic classes of similarity functions among categories.

1. *Boolean similarity function.* This type of function can only distinguish whether a given concept is present or not. Though simple, it is used in some cases.

2. *Real-valued similarity function.* This type of function assigns real values in the interval [0, 1] to pairs of concepts and enables a rich structuring of the concept space.

From the similarity functions for single concepts, in general, similarity functions for concept sets are derived. For the case of boolean similarity function, this may result in more complex similarity measures for concept sets.

Examples of concept spaces. In the following, we provide some frequent examples of concept spaces and their similarity measures that have been proposed for the construction of semantic overlay networks.

When using flat concept spaces [Michlmayr et al., 2007, Tirado et al., 2010], the similarity among two concepts c_1 and c_2 degenerates to the equality function, such that $sim(c_1, c_2) = 1$ iff. $c_1 = c_2$, i.e., a Boolean similarity function. For sets of concepts, related, for example, to collections of queries or resources, a profile can be derived by constructing the frequency distribution of the concepts in the sets. Similarity among concept sets can then be computed similarly as similarity for term frequency vectors in full text retrieval using the cosine similarity measure.

Hierarchical categorization has been widely used [Crespo and Garcia-Molina, 2005, Haase et al., 2004, Penzo et al., 2008, Schmitz, 2004]. The hierarchies are taken from domain-specific ontologies, e.g., the ACM topic hierarchy, generic ontologies, such as Wordnet, or one of the abundant hierarchical categorizations found on portal sites on the Web. With hierarchical categorization, the problem of determining the similarity among two concepts becomes more complex. Sophisticated methods are used to define similarity of concepts by exploiting the hierarchical structure of the concept space. Examples of such measures are

- *Shortest path distance.* The similarity between two concepts c_1 and c_2 is defined as the length s of the shortest path connecting them in the concept space, i.e., $sim(c_1, c_2) = \frac{s}{s_{max}}$, where s_{max} is the maximal path length in the concept hierarchy.

- *Weighted path distance.* This measure considers in addition to the shortest path length s also the height l of the common ancestor of the two concepts, resulting in the following definition [Haase et al., 2004].

$$sim(c_1, c_2) = \begin{cases} e^{-\alpha s} \frac{e^{\beta l} - e^{-\beta l}}{e^{\beta l} + e^{-\beta l}}, & \text{if } c_1 \neq c_2, \\ 1 & \text{otherwise} \end{cases}$$

- *GM distance.* This measure relies only on the height of the concepts in the hierarchy, where $l(c)$ is the height of a concept c in the hierarchy and l the height of the common ancestor [Penzo et al., 2008].

$$sim(c_1, c_2) = \begin{cases} \frac{2l}{l(c_1) + l(c_2)}, & \text{if } c_1 \neq c_2, \\ 1 & \text{otherwise} \end{cases}$$

Again, these similarity measures can be generalized to sets of concepts. For example, for comparing the set of concepts related to a query Q with the set of concepts related to a peer profile R, the similarity can be given as [Haase et al., 2004]

$$Sim(Q, R) = \frac{1}{|Q|} \sum_{c \in Q} max_{c' \in R} sim(c, c') \ .$$

For comparing profiles of different peers, the following similarity function can be applied [Raftopoulou and Petrakis, 2008].

$$Sim(R_1, R_2) = max_{c_1 \in R_1, c_2 \in R_2} sim(c_1, c_2) \ .$$

Concepts extracted from text or media files are usually represented as vectors in high-dimensional feature spaces [Doulkeridis et al., 2007, Falchi et al., 2007, Haghani et al., 2009, Li et al., 2004, Parreira et al., 2007, Raftopoulou and Petrakis, 2008, Siebes and Kotoulas, 2007, Tang et al., 2003]. Similarity among concepts is then computed in the simplest case as the cosine similarity among the feature vectors. More sophisticated similarity measures are found in particular for text retrieval, such as generalized tf-idf measures or similarity measures based on language models [Zhai, 2008].

One problem with feature vectors is their high dimensionality. Distance-based clustering and routing in a high-dimensional space, as realized by small world overlay networks, becomes ineffective in such a case due to the so-called *curse of dimensionality* [Weber et al., 1998]. Therefore, methods for reducing the dimensionality of the high-dimensional feature space to a lower dimensional concept space are used. *Latent semantic indexing* [Deerwester et al., 1990] is one method that is applied to that end [Li et al., 2004, Siebes and Kotoulas, 2007, Tang et al., 2003]. Applied to full text retrieval, it maps text vectors to *concept vectors* of dimension typically 50 to 300. Similarity of concept vectors is as for feature vectors computed using cosine similarity.

Latent semantic indexing. In the vector space model, the relation between the terms T and the documents D can be represented as a $|T| \times |D|$ matrix M, whose entries are the term weights from a tf-idf weighting scheme, which indicate the importance of a term for the document. Assume the matrix M has rank r. Then it can be decomposed using a *singular value decomposition* into three matrices as follows $M = U \cdot \Sigma \cdot V^T$. $U = (u_1, \ldots, u_r)$ is an orthonormal $|T| \times r$ matrix, $\Sigma = diag(\sigma_1, \ldots, \sigma_r)$ is a diagonal $r \times r$ matrix with $\sigma_1 \geq \ldots \geq \sigma_r$, and $V = (v_1, \ldots, v_r)$ is an orthonormal $|D| \times r$ matrix. In order to approximate the matrix M, only the s largest singular values $\sigma_1, \ldots, \sigma_s$ are retained. This results in an approximation $M_s = U_s \cdot \Sigma_s \cdot V_s^T$, with $U_s = (u_1, \ldots, u_s)$, $V_s = (v_1, \ldots, v_s)$ and $\Sigma_s = diag(\sigma_1, \ldots, \sigma_s)$. The rows of $V_s \cdot \Sigma_s$ correspond to the documents, and the columns can be interpreted as the concept dimensions of a latent semantic space. Similarity of documents can be computed using cosine similarity. Similarly, the rows of $U_s \cdot \Sigma_s$ correspond to the terms of the vocabulary.

Sets of related concepts represented as feature vectors form *clusters* in the feature space. In the simplest case, they can be represented by their *centroid vector*, and the similarity of clusters can be computed again using cosine similarity [Raftopoulou and Petrakis, 2008].

5.2 SEMANTIC OVERLAY NETWORK CONSTRUCTION

The approach to constructing a semantic overlay network is to a certain extent orthogonal to the choice of the concept space and its similarity metrics. We provide, first, a general classifications of possible approaches and illustrate those then by concrete examples.

A first distinction can be made with respect to the criterion used to establish links between peers that are semantically close, i.e., the *semantic clustering* strategy. Three main approaches can be distinguished here

1. *Interest-Resource clustering.* Peers create preferably links to peers that hold resources of interest to them, typically to peers that have provided useful answers to queries earlier. As a result, peers can more easily find relevant resources and peers with common interest profiles tend to cluster around peers holding related resources.

2. *Resource-Resource clustering.* Peers create preferably links to peers that hold similar resources. As a result, peers with similar types of resources tend to cluster and access to resources of a specific type is localized and thus more efficient.

3. *Interest-Interest clustering.* Peers create preferably links to peers that have requested similar types of resources the peer is interested in. In this way, peers obtain links to *recommenders* that might be particularly knowledgeable where to locate specific types of resources.

A second distinction can be made with respect to the mechanism that is used to create the semantic overlay network. Here we can distinguish two main classes of approaches:

1. *Protocol-driven overlay network creation.* The connectivity of an existing, typically unstructured, overlay network is augmented or modified through protocols that implement preferential attachment to peers with related interests or resources. In this way, initially unstructured overlay networks are gradually transformed into increasingly structured overlay networks that cluster semantically related peers. We find three approaches of how the structure of routing tables in existing overlay networks is modified in this approach.

   (a) *Annotation.* A peer annotates the links in its routing table with information on what type of content can be found when following the link.

   (b) *Augmentation.* A peer maintains a ranked list of peers that have specific relevance for a peer's interest profile. This list is often called *shortcut list*.

   (c) *Modification.* A peer modifies the structure of the existing overlay network in order to better connect to peers with similar interests and profiles.

   With protocol-driven semantic overlay network creation the structure of the semantic overlay network is *implicitly* specified through the network creation protocol. Consequently, with this approach, structural properties or performance guarantees are often not available.

2. *Mapping to a structured overlay network.* The concept space is mapped to the identifier space of a structured overlay network, by preserving the proximity of semantically close concepts. This assures that semantically related peers are clustered with the structured overlay network. With this approach, the structure of the semantic overlay network is *explicitly* specified and structural properties and performance characteristics can be given.

5.3 PROTOCOL-DRIVEN SON

Protocol-driven approaches to semantic overlay construction use either existing messages, such as queries, to obtain information about semantically related peers, or use specific messages to announce or search for such information proactively. We now provide examples of several approaches that combine the aforementioned design dimensions in different ways.

Shortcuts lists. A basic approach for clustering peers with common interests is to adapt the routing tables in an overlay network by introducing new links, so-called *shortcuts*, to peers that successfully answered a query. This approach has been proposed for unstructured overlay networks, such as Gnutella [Sripanidkulchai et al., 2003]. Peers maintain a shortcut list of bounded length to those peers that have successfully provided answers to earlier queries. The peers in the shortcut list are ranked based on different criteria, such as the probability of providing relevant content in earlier queries, the amount of content available at the peer, but also technical factors related to latency, bandwidth or workload. Before processing a query using the standard overlay network mechanism, such as flooding, the peer first contacts the members of its shortcut list starting from the top.

Experimental evaluations show that significant message load reduction can be achieved in the peer-to-peer network. Note that in this approach, the concept space is not made explicit, but it is only implicitly reflected in the interest relationships among peers.

This approach can be refined by explicitly modeling the type of information a peer has successfully provided in earlier interactions [Löser et al., 2005, Tempich et al., 2004]. Hierarchically organized categories are associated with the queries. For example, in RDF, such categories can be extracted from the subjects or objects occurring in atomic triple queries. Peers can now identify other peers that match their interests in two ways. First, they can find *content providers,* which are peers that have provided successfully responses to a query related to a category. Second, by overhearing query traffic that they forward, they can find *recommenders*, which are peers that pose queries on specific categories, and thus might be peers that are well connected with content providers for these categories. The peers maintain a shortcut list of the form $shortcut(c, p, r, t)$, where c is a category, p a peer, r indicates of whether it is a recommender ($r = 1$) or content provider ($r = 0$) and t is the time of the last update. Entries in the short cut list are ranked according to a relevance function, such as [Löser et al., 2005]

$$rel(c, p, r, t) = \alpha\ max_{c' \in C_p} sim(c, c') + \beta(1 - \frac{t}{2}) + (1 - \alpha - \beta)\frac{t - t_{min}}{t_{min}}$$

where C_p is the set of categories locally stored at peer p, $0 < \alpha, \beta < 1$ and t_{min} the oldest update time in the shortcut list. The elements with lowest relevance are evicted when the shortcut list overflows. When processing a query, the k elements with the highest relevance are chosen for query forwarding. If less than k elements are found, some randomly selected peers are added. If k relevant elements are found, a few are replaced with random peers to avoid over-fitting of the network structure. Experimental evaluations show that using category-specific shortcuts helps to further reduce message load in the peer-to-peer network. Other approaches to perform category specific interest clustering have been based on stigmetric algorithms [Michlmayr et al., 2007].

Routing indices. The concept of *routing indices* has been introduced to enhance in unstructured overlay networks existing routing tables with information on which links connect to parts of the network with documents pertaining to a specific topic [Crespo and Garcia-Molina, 2002]. In this approach, peers proactively advertise their resources. A flat category space C is given, and for each neighbor $p_n \in N(p)$, a statistics $f_c(p_n)$ is maintained on the number of resources can be found, following the link to the neighboring peer p_n and that belong to each of the categories $c \in C$. Also the total number $f_{tot}(p_n)$ of documents reachable over the link to p_n is maintained. When processing a conjunctive query Q with multiple categories, the relevance of each link to a neighbor p_n is determined as

$$rel(Q, p_n) = f_{tot}(p_n) \prod_{c \in Q} \frac{f_c(p_n)}{f_{tot}(p_n)}$$

The query is forwarded to the neighbors in order of their relevance till a sufficient number of results is obtained. Thus, this approach prefers connections to peers with relevant resources. The approach can be further refined by maintaining a statistics $f_c(p_n, l)$ that provides the information how many documents for category c can be reached in l hops when following the link to p_n and making relevance dependent on the distance. When a peer joins the network or has significant changes in its document collection, it updates its neighbors with the new statistics, who in turn further propagate the updates to their neighbors. The main challenge of this approach is to maintain the statistics in the presence of cycles in the network, e.g., by using techniques for aggregation as have been discussed in Section 2.2.3.

Advertisements. Advertisements of peers can be used to change the overlay network structure. This can be achieved in two ways. Peers can broadcast a message containing their resource profile, represented as concept set or concept vector, in the peer-to-peer network [Haase et al., 2004, Siebes and Kotoulas, 2007]. Other peers receiving such advertisements retain them, when the advertisement is sufficiently similar to the peer's own profile. Alternatively, peers can also proactively search for semantically related peers, e.g., by performing a random walk [Raftopoulou and Petrakis, 2008, Schmitz, 2004]. A random walker collects the profiles of peers it visits and returns them to the originator of the random walk. Then the originator selects the most similar peers and connects to them, while abandoning existing connections to less similar peers. Instead of visiting only randomly selected peers through a random walk one could also select peers for the next step according to the highest profile similarity. This results in a *gradient walk* [Raftopoulou and Petrakis, 2008].

A similar idea is to connect to semantically related peers by preferentially contacting the neighbors of the peer's own neighbors. In this approach, each peer p maintains a list of neighbors $N(p)$ and a list of candidates $C(p)$, which are potential neighbors. A peer executes occasionally Algorithm 8 with parameters $0 < \alpha < \beta < 1$ and similarity threshold $0 < \tau < 1$ [Parreira et al., 2007].

Other approaches for clustering peers in the concept space have been making explicit use of clustering algorithms [Doulkeridis et al., 2007, Penzo et al., 2008].

Query processing. When processing a query in a semantic overlay network in which semantically related peers have been clustered, different query processing strategies can be applied [Schmitz, 2004]. We assume a query is forwarded in a unstructured overlay network with a limited time-to-live. Then possible forwarding strategies are the following:

1. *Random forwarding.* The query is forwarded to k randomly selected peers.

2. *Fixed fanout forwarding.* The query is forwarded to the k peers whose profile is most similar to the query profiles.

3. *Threshold forwarding.* The query is forwarded to all peers whose similarity is above a given threshold τ.

4. *Fireworks.* If the similarity between the query and the peers profile is above a given threshold τ, a new broadcast with limited time-to-live is initiated at the peer.

Algorithm 8: $findNeighbor(p)$

```
1  r = random(0, 1);
2  if r < α then
3  |   p contacts a neighbor p_n ∈ N(p);
4  |   if p_n is alive then
5  |   |   C(p) = C(p) ∪ N(p_n)
6  |   else
7  |   |   N(p) = N(p) \ {p_n}
8  |   end
9  else
10 |   if r < β then
11 |   |   p contacts a candidate p_c ∈ C(p)
12 |   else
13 |   |   p contacts a randomly selected peer p_c
14 |   end
15 |   if p_c is alive and sim(p, p_c) > τ then
16 |   |   N(p) = N(p) ∪ {p_c};
17 |   |   evict from N(p) the peer with the lowest similarity
18 |   end
19 end
```

5.4 MAPPING-BASED SON

Mapping-based approaches to semantic overlay construction provide an explicit mapping of a concept space to the identifier space of a structured overlay network. This mapping preserves the similarity relationship in the concept space. The approach is mostly used to cluster peers with similar resources, thus for resource-resource clustering. The main challenge with this approach is the discrepancy between the dimensionality of concept spaces, which is usually high, and the dimensionality of identifier spaces of structured overlay networks, which is usually low. Different techniques have been proposed to overcome this dimensionality gap, some of which we describe in the following.

5.4.1 MAPPING CATEGORIES

The following approach provides a mapping from a category space into a one-dimensional structured overlay network such as Chord [Tirado et al., 2010]. Resources are classified by associating with them one or more categories from a predefined category set $C = \{c_1, \ldots, c_k\}$. Peers are associated with the category for which they have the largest number of documents. The peer identifier for the overlay network is generated by hashing the concatenation of the category identifier i of category c_i it is associated with and the physical address a of the peer, i.e., $id = h(i|a)$. In this way, peers from the same category will be mapped into consecutive segments of the overlay identifier space. Since resources can be associated with several categories, for each category, one can determine the strength of correlation with other categories by analyzing the set of resources held by peers in that category. Let $f(c_p, c_r)$ be the number of resources associated with category c_r that are found at a peer associated with category c_p. Then the *affinity* of two categories c_i and c_j can be defined as

$$a(c_i, c_j) = \frac{f(c_i, c_j) + f(c_i, c_j)}{\sum_{l=1...k}(f(c_i, c_l) + f(c_l, c_i))}.$$

Note that affinity is an asymmetric measure. The affinity matrix for categories is given as $A = [a(c_i, c_j)]_{i,j=1...k}$. The affinity matrix is known to all peers and can be approximately maintained by methods for data aggregation such as those discussed in Section 2.2.3. It is used in order to determine an ordering of the clusters in the one-dimensional identifier space that maximizes locality. To that end, the following clustering algorithm is applied to A:

Algorithm 9: $clusterCategories(A)$

```
1 select (r, c) such that A_rc = max_{i,j=1...k} A_ij;
2 π(1) = r;
3 for l = 2 to k do
4 |   π(l) = c;
5 |   r := c;
6 |   select c such that A_rc = max_{j=1...k, j∉{π(1),...,π(l)}} A_rj;
7 end
```

The permutation π is then applied to the category set to determine the optimal ordering of the categories in the overlay network identifier space. Searches for queries consisting of one or more categories can then be evaluated by using range search techniques that have been introduced in Section 2.1.5.

5.4.2 MAPPING CONCEPT VECTORS

The problem of mapping a concept space to a structured overlay network has also been studied for concept vectors, such as those generated by latent semantic indexing. The term frequency statistics for constructing the LSI mapping is assumed to be available from a document sample. It could also be obtained in the peer-to-peer network using aggregation techniques.

Mapping to a lower-dimensional space. Since concept vectors are multi-dimensional, a natural approach is to map them to a multi-dimensional content-addressable network, such as CAN [Tang et al., 2003]. In the simplest case, the multi-dimensional concept space is directly used as identifier space for the overlay network. However, this approach suffers from poor search performance, since the dimensionality of the LSI space is considerably higher (50 to 300) than what is needed for efficient routing in content-addressable networks. The optimal dimension of the key space depends, in particular, on the size of the peer-to-peer network. E.g., for a network of size n a dimension of $m = 2.3 \cdot \log(n)$ is suggested, which results in values of approximately $m = 30$ for peer-to-peer networks of a typical size $n = 10^6$. Therefore, a more sophisticated mapping is required.

Given a structured overlay network with an identifier space of dimensionality m and a concept space with dimensionality $d = l \cdot m$. Then a concept vector $\vec{v}$ can be partitioned into vectors $\vec{v}_i = (v_{(i-1)\cdot m+1}, \ldots, v_{i\cdot m}), i = 1, \ldots, l$. These l vectors are used as identifiers for indexing the document in the overlay network, i.e., the document is indexed l times. For querying, similarly, the query vector is partitioned, and the documents found in the vicinity of those identifiers are retrieved from the overlay network. Finally, the set of documents that are similar for the complete key is selected from the document set retrieved in this way.

When using concept spaces which have a metric similarity function, but of which the structure and dimensionality are not known, a mapping to a low-dimensional identifier space of dimension m can be constructed by using *pivots* [Falchi et al., 2007]. m elements $c_1, \ldots, c_m \in C$ of the original set of concept vectors C are chosen. For a resource r, the key is then generated by computing the m-dimensional vector

$$(sim(r, c_1), \ldots, sim(r, c_m)) .$$

The pivot elements can be chosen uniformly at random from C. More sophisticated methods for optimized pivot selection have been proposed in the literature [Bustos et al., 2003]. The m-dimensional representations of the concept vectors are then managed in a content-addressable network.

An alternative method for mapping to lower-dimensional spaces is to use *locality-sensitive hash functions* [Gionis et al., 1999]. The salient property of these hash functions is that similar objects are more likely mapped to the same hash value than distant ones. More precisely, a hash function $h : C \rightarrow I$ is $(\rho_1, \rho_2, \tau_1, \tau_2)$-locality sensitive if for $c_1, c_2 \in C$ the following conditions are met:

- if $sim(c_1, c_2) \leq \rho_1$ then $Pr[h(c_1) = h(c_2)] \geq \tau_1$

- if $sim(c_1, c_2) > \rho_2$ then $Pr[h(c_1) = h(c_2)] \leq \tau_2$

More generally, it has been shown that with these functions objects that are close in the concept space are more likely to be mapped to hash values that are close, thus maintaining locality [Haghani et al., 2009].

Locality-sensitive hash functions. A special case of a locality sensitive hash (LSH) function is the family of LSH, which are based on p-stable distributions. We describe the case of $p = 2$. For a d-dimensional data point $c \in C$, the hash value is given for real number w as

$$h_{a,b} = \lfloor \frac{a \cdot c + b}{w} \rfloor$$

where

- a is a d-dimensional random vector whose elements are drawn from a normal distribution $N(0, 1)$,
- b is a real number drawn uniformly at random from $[0, w]$.

Using m such hash functions, a data point can be mapped to a m-dimensional integer vector.

Mapping to a one-dimensional space. Another method to reduce the dimensionality of the concept space is to arrange the subspaces of the concept space the peers are associated with in a linear order and to map them to a structure overlay network using an order-preserving hash function. This avoids multiple indexing and searching for one concept vector when partitioning of concept vectors is performed as described before. The principle of this approach is similar to the one used for mapping multi-dimensional attribute spaces to one-dimensional identifier spaces, as discussed earlier in Section 2.2.1. However, since concept vectors are non-uniformly distributed, space-filling curves such as Hilbert curves are not suitable for that purpose. Thus, the linear order needs to be dynamically maintained when peers are splitting or joining subspaces during joins and departures [Li et al., 2004].

The idea of using pivot elements can be used to map high-dimensional concept vectors to a one-dimensional structured overlay network [Sahin et al., 2005]. A set of pivot $c_1, \ldots, c_m \in C$ is chosen, and then for each resource r, the similarities with the pivots are computed. The k pivot elements with the highest similarities are selected, and their identifiers are concatenated in order of decreasing similarity to generate a resource key. This key is then hashed to the identifier space of the overlay network. In this way, similar resources tend to be clustered in the overlay network.

The idea of using locality-sensitive hash functions can be used to construct another mapping to a one-dimensional identifier space of an overlay network [Haghani et al., 2009]. Given m locality-sensitive hash functions $h_{a_1,b_1} \ldots, h_{a_m,b_m}$, a concept vector c can be mapped first to an m-dimensional integer vector

$$g(c) = (h_{a_1,b_1}(c), \ldots, h_{a_m,b_m}(c)) .$$

This vector is further mapped to an integer $id(g(c))$ by summation, i.e.,

$$id(g(c)) = \sum_{i=1}^{m} g_i(c) .$$

Using summation has two properties. First, locality is maintained since for similar integer vectors also the sums will be similar. Second, the distribution of identifiers generated by this approach is known. Since a is generated using a normal distribution $N(0, 1)$, $a \cdot c$ is distributed following the Normal distribution $N(0, ||c||_2)$. For small w then h_{a_i,b_i} is distributed according to $N(\frac{1}{2}, \frac{||c||_2}{w})$, and therefore $id(g(c))$ is distributed according to $N(\frac{k}{2}, \frac{\sqrt{k}||c||_2}{w}$. The global distribution of identifiers generated from a concept set C is then

$$N(\frac{k}{2}, \frac{\sqrt{k \sum_{c \in C} ||c||_2}}{w\sqrt{|C|}}) .$$

Since this is a static distribution and it can be approximately known, methods for static range partitioning, as introduced in Section 2.1.1, can be applied to ensure load balancing in the one-dimensional overlay network.

Conclusion. We have seen in this Section and in the earlier Section 2.2.1 on multiple attribute management that there exists a wide variety of approaches to map from higher-dimensional feature and attribute spaces to lower-dimensional or one-dimensional identifier spaces for structured overlay networks. Figure 5.2 provides an overview of the approaches we have introduced in this book. But it is important to realize that this is only a small selection of methods that have been considered in the context of peer-to-peer data management, and many other methods for dimensionality reduction developed in other fields might be as well applicable.

The high attention that this problem receives may be well be related to the fact that many complex networks, both natural and engineered networks, with efficient routing are based on a hidden metric space [Boguna et al., 2009]. With semantic overlay networks, we know about the nature of the underlying metric space. It is the space of semantic concepts. We take advantage of this knowledge by either exploring this space using communication protocols, as shown in Section 5.3, or by explicitly exploiting its structure for the construction of a routing efficient and locality preserving overlay network, as shown in the previous Section 5.4.

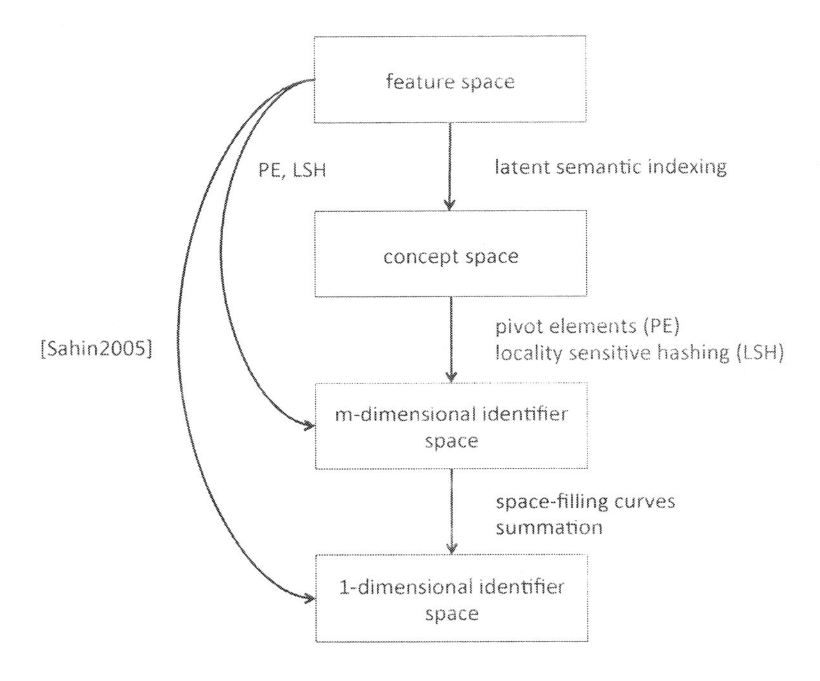

Figure 5.2: Overview of approaches to dimensionality reduction for semantic overlay network construction.

CHAPTER 6

Conclusion

We have limited the scope of this lecture on purpose to problems related to representation and search of data and documents in peer-to-peer systems in order to maintain a clear focus. In particular, we have not covered problems related to state management and data dissemination as well as some more specialized techniques of data search and analysis. Some of these areas have received relatively small attention in the literature (e.g., transactions), some relate to functions that are typically handled at the peer-to-peer overlay network layer (e.g., replication for resilience), and some are beyond the scope of the topic of peer-to-peer data management though they rely on peer-to-peer data management services (e.g., trust management and social networks, content distribution).

In the following, we shortly survey topics not covered in this lecture and provide some pointers to the literature for further reading. Some of these topics might become subject of future editions of this lecture.

Specialized queries. Apart from the basic queries, essentially corresponding to the expressivity of SQL or Sparql, processing techniques for a number of non-standard query types have been studied in the literature. This concerns in particular queries that exploit similarity relationships, such as top-k [Balke et al., 2005], nearest neighbor [Haghani et al., 2009] and skyline queries [Wang et al., 2007].

Data mining. As peer-to-peer data management systems typically provide access to very large data collections, it is a natural to also apply data mining techniques within such systems to uncover hidden correlations and structures relevant to applications. Some data mining techniques for which peer-to-peer algorithms have been devised are association rule mining, clustering and classification [Datta et al., 2006]. Some of the techniques we have introduced in this lecture, such as computing aggregates, latent semantic indexing or clustering in semantic overlay networks are approaches in the spirit of or in need of data mining algorithms.

Replication. In order to increase availability and durability of data in peer-to-peer systems, replication and erasure coding techniques have been proposed [Datta, 2009]. Another use of replication is to improve response time in case of skewed query workloads. Essentially, data items that are more popular are replicated more frequently and thus can be served by more peers. These approaches have been considered both when managing data in unstructured and structured overlay networks. Since most peer-to-peer overlay networks also include a mechanism for replication, an interesting question is at which layer these mechanisms should be provided, i.e., the overlay network layer or the peer-to-peer data management layer.

Updates. Closely related to the problem of replication is the problem of maintaining consistency of replicas under updates [Ives, 2009]. In the simplest case, copies have a well-defined owner, who is the only one allowed to perform updates or resolves conflicts among different updates. In order to avoid having a single owner becoming a bottleneck in maintaining a data object, more advanced approaches use consensus protocols or allow for partial divergence of replicas.

Transactions. For applications with stronger consistency requirements, transaction mechanisms have been studied also in the context of peer-to-peer data management [Masud and Kiringa, 2011]. Some of these are adaptations of distributed transaction protocols well known from the database literature that take into account the decentralized nature of a peer-to-peer architecture.

Mobile data management. Peer-to-peer data management bears many similarities with data management in mobile ad-hoc networks [Schollmeier et al., 2002]. Both types of systems are based on decentralized architectures and self-organization and use an underlay network to implement higher level data management tasks. Key differences are that network connectivity in mobile networks is much more volatile due to node mobility and wireless connections, which requires special precautions, e.g., caching of data in case of disconnection, and that the network structure is constrained by the physical location of the devices.

Stream data management. Data stream processing is an area that over the past years has been attracting growing attention due to the increasing number of real-time data sources on the Web and from sensor networks. Distributed data stream processing systems have been typically taking a peer-to-peer architectural approach, in order to ensure scalability with a growing number of stream processing nodes. Examples of such systems are Medusa [Cherniack et al., 2003], DSMI [Kumar et al., 2005], StreamGlobe [Kuntschke et al., 2005] and Global Sensor Networks [Aberer et al., 2007].

Publish-subscribe and continuous queries. When the number of information producers and consumers grows and information is continuously generated, the search paradigm shifts from information pull to information push. In the publish-subscribe model, information consumers subscribe to information sources, qualifying their needs, and sources push matching data over the network to their subscribers. Due to the inherently distributed nature and large scale of such systems, peer-to-peer architectures are a natural approach to support their efficient implementation [Li, 2008, Triantafillou and Aekaterinidis, 2009]. Some of the approaches to process aggregation queries introduced in this lecture can be understood as special cases of such an approach.

Content distribution. Today, the most popular use of peer-to-peer network is for content distribution, e.g., with Bittorrent. The approach these systems take is to split very large content files, such as videos, into small parts that are then quickly replicated to a large number of clients using swarming techniques. Techniques for speeding up the distribution of content by exploiting the bandwidth at clients is an active area of research [Felber and Biersack, 2009]. In addition, content distribution systems are in need for search mechanisms to locate contents. These can be realized using peer-to-peer data management techniques as introduced in this lecture.

Trust management. Trust and reputation are central concepts in making interactions among participants in large-scale systems more reliable when frequently strangers are encountered [Josang et al., 2007]. Therefore, this concept has also been applied from the early days to peer-to-peer systems [Despotovic, 2009]. In particular, managing the reputation data required by trust management systems poses genuine technical problems for peer-to-peer data management [Aberer and Despotovic, 2001].

Access control, privacy, and security. Due to their decentralized nature, peer-to-peer systems are particularly exposed to number of possible attacks that can compromise their security [Wallach, 2003]. Trust management systems, as mentioned above can avoid some potential problems, but also classical security mechanisms, such as access control [Crispo et al., 2005] and privacy preservation [Clarke et al., 2001, Dingledine et al., 2001], have been considered for that purpose.

Social networks. Recently social networks have had a tremendous success and wide adoption, beyond what peer-to-peer networks, which can be considered as a predecessor of them, have ever achieved. This poses the question whether they might also serve as substrate for data management. On the one hand, this appears appealing as these systems handle significant amounts of data, and on the other hand, it might be a way to address privacy concerns with centralized solutions under the control of a few providers as well as trust concerns in open peer-to-peer architectures. For example, developments in this direction have started recently in the area of peer-to-peer systems for content distribution leveraging on social networks [Galuba et al., 2010, Pouwelse et al., 2008].

APPENDIX A

Notational Conventions

The notational conventions that have been used in the book are summarized in the following tables. First, we summarize notations for sets, both corresponding to domains and declarative specifications. Note that we accept overloading of symbols if their meaning is related or no confusion can occur.

A, B	attribute sets
C	logical conditions
	classes
D	documents
	data space
	database instance
F	features
I	identifier space of a peer-to-peer system
K	set of keywords
L	posting list
M	mapping tables
	mapping paths
$N(p)$	neighbors of a peer p
P	peers in a peer-to-peer system
R	resources managed in a peer-to-peer system
$R(a_1, \ldots, a_k)$	a relation with attributes $a_1, \ldots, a_k$
T	term vocabulary
V	domain of attribute values
Q	query
S	schema

Next, we summarize notations used for elements of domains.

$d \in D$	document
$id \in I$	an identifier
$f \in F$	feature
$m \in M$	schema mappings
$p \in P$	individual peers
$r \in R$	a resource
$t \in R$	tuple of a relation
$t \in T$	term
$v \in V$	value from an attribute domain

Notations for various characteristics quantities of a peer-to-peer data management system are the following.

e	number of edges in a peer-to-peer network
f	frequency
k	dimensionality of multi-dimensional attribute space number of relations
l	document length length of a bitstring level in a binary tree number of conditions
m	dimensionality of multi-dimensional identifier space
n	the number of peers in a peer-to-peer system
s	size of a sample size of a document collection

The following are notations for functions.

$d : I \times I \rightarrow \mathbb{R}$	distance function in the identifier space
$deg(p) : P \rightarrow \mathbb{N}$	in-degree of a peer
h	a hash function
$h_P : P \rightarrow I$	the hash function to map peers into identifiers
$h_R : R \rightarrow I$	the hash function to map resources into identifiers
$\mu : V^n \rightarrow V$	aggregation function

Finally, the following notations are used for probabilistic measures and various thresholds.

Pr	probability
E	expectation value
Var	variance
λ	probability value
σ	variance
δ	granularity
τ	threshold
ε	error

Bibliography

Karl Aberer. P-grid: A self-organizing access structure for p2p information systems. In *Proc. Int. Conf. on Cooperative Information Systems*, pages 179–194. Springer Berlin / Heidelberg, 2001. DOI: 10.1007/3-540-44751-2_15 6, 7

Karl Aberer and Zoran Despotovic. Managing trust in a peer-2-peer information system. In *Proc. Int. Conf. on Information and Knowledge Management*, pages 310–317, 2001. DOI: 10.1145/502585.502638 111

Karl Aberer and Manfred Hauswirth. An overview of peer-to-peer information systems. In *4th International Meeting on Distributed Data & Structures*, volume 14 of *Proceedings in Informatics*, pages 171–188. Carleton Scientific, 2002. 1

Karl Aberer, Philippe Cudre-Mauroux, and Manfred Hauswirth. A framework for semantic gossiping. *ACM SIGMOD Rec.*, 31:48–53, 2002a. DOI: 10.1145/637411.637419 57

Karl Aberer, Magdalena Punceva, Manfred Hauswirth, and Roman Schmidt. Improving data access in p2p systems. *IEEE Internet Comput.*, 6:58–67, 2002b. DOI: 10.1109/4236.978370 7, 21, 30

Karl Aberer, Philippe Cudre-Mauroux, and Manfred Hauswirth. The chatty web: emergent semantics through gossiping. In *Proc. 12th Int. World Wide Web Conf.*, pages 197–206, 2003. DOI: 10.1145/775152.775180 57, 68

Karl Aberer, Luc Onana Alima, Ali Ghodsi, Sarunas Girdzijauskas, Seif Haridi, and Manfred Hauswirth. The essence of p2p: A reference architecture for overlay networks. In *Proc. 5th IEEE Int. Conf. on Peer-to-Peer Computing*, pages 11–20, Los Alamitos, CA, USA, 2005. DOI: 10.1109/P2P.2005.38 1

Karl Aberer, Manfred Hauswirth, and Ali Salehi. Infrastructure for data processing in large-scale interconnected sensor networks. In *Proc. 2007 Int. Conf. on Mobile Data Management*, pages 198–205, 2007. DOI: 10.1109/MDM.2007.36 110

Serge Abiteboul and Oliver M. Duschka. Complexity of answering queries using materialized views. In *Proc. 17th ACM SIGACT-SIGMOD-SIGART Symp. on Principles of Database Systems*, pages 254–263, New York, NY, USA, 1998. ACM. DOI: 10.1145/275487.275516 63

Serge Abiteboul, Ioana Manolescu, Neoklis Polyzotis, Nicoleta Preda, and Chong Sun. Xml processing in dht networks. In *Proc. 24th Int. Conf. on Data Engineering*, pages 606–615, 2008. DOI: 10.1109/ICDE.2008.4497469 81, 89, 94

Keno Albrecht, Ruedi Arnold, Michael Gahwiler, and Roger Wattenhofer. Aggregating information in peer-to-peer systems for improved join and leave. In *Proc. 4th IEEE Int. Conf. on Peer-to-Peer Computing*, pages 227–234. IEEE Computer Society, 2004. DOI: 10.1109/PTP.2004.1334951 45

Stephanos Androutsellis-Theotokis and Diomidis Spinellis. A survey of peer-to-peer content distribution technologies. *ACM Comput. Surv.*, 36:335–371, 2004. DOI: 10.1145/1041680.1041681 1

Artur Andrzejak and Zhichen Xu. Scalable, efficient range queries for grid information services. In *Proc. 2nd IEEE Int. Conf. on Peer-to-Peer Computing*, pages 33–40, Los Alamitos, CA, USA, 2002. DOI: 10.1109/PTP.2002.1046310 34

Nick Antonopoulos and Lee Gillam. *Cloud Computing - Principles, Systems and Applications*. Springer, 1st edition, 2010. 13

Benjamin Arai, Gautam Das, Dimitrios Gunopulos, and Vana Kalogeraki. Approximating aggregation queries in peer-to-peer networks. In *Proc. 22nd Int. Conf. on Data Engineering*, page 42, Los Alamitos, CA, USA, 2006. DOI: 10.1109/ICDE.2006.23 46, 47

Wolf-Tilo Balke, Wolfgang Nejdl, Wolf Siberski, and Uwe Thaden. Progressive distributed top-k retrieval in peer-to-peer networks. In *Proc. 21st Int. Conf. on Data Engineering*, volume 0, pages 174–185, 2005. DOI: 10.1109/ICDE.2005.115 109

Farnoush Banaei-Kashani and Cyrus Shahabi. Swam: a family of access methods for similarity-search in peer-to-peer data networks. In *Proc. Int. Conf. on Information and Knowledge Management*, pages 304–313, New York, NY, USA, 2004. DOI: 10.1145/1031171.1031236 38

Carlos Baquero, Paulo Sergio Almeida, and Raquel Menezes. Fast estimation of aggregates in unstructured networks. In *5th Int. Conf. on Autonomic and Autonomous Systems*, pages 88–93, 2009. DOI: 10.1109/ICAS.2009.31 44

Dominic Battre. Caching of intermediate results in dht-based rdf stores. *Int. J. Metadata Semant. Ontologies*, 3:84–93, 2008. DOI: 10.1504/IJMSO.2008.021207 53

Mayank Bawa, Hector Garcia-Molina, Aristides Gionis, and Rajeev Motwani. Estimating aggregates on a peer-to-peer network. Technical report, Computer Science Department, Stanford University, 2003. 42, 43

Matthias Bender, Sebastian Michel, Peter Triantafillou, Gerhard Weikum, and Christian Zimmer. Improving collection selection with overlap awareness in p2p search engines. In *Proc. 31st Annual Int. ACM SIGIR Conf. on Research and Development in Information Retrieval*, pages 67–74, 2005. DOI: 10.1145/1076034.1076049 86

Jon Louis Bentley. Multidimensional binary search trees used for associative searching. *Commun. ACM*, 18(9):509–517, 1975. DOI: 10.1145/361002.361007 38

Ashwin R. Bharambe, Mukesh Agrawal, and Srinivasan Seshan. Mercury: supporting scalable multi-attribute range queries. In *Proc. ACM Int. Conf. on Data Communication*, pages 353–366, New York, NY, USA, 2004. DOI: 10.1145/1015467.1015507 28, 37

Christian Bizer. The emerging web of linked data. *IEEE Intelligent Systems*, 24(5):87–92, 2009. DOI: 10.1109/MIS.2009.102 51

Burton H. Bloom. Space/time trade-offs in hash coding with allowable errors. *Commun. ACM*, 13: 422–426, 1970. DOI: 10.1145/362686.362692 80

Marian Boguna, Dmitri Krioukov, and K. C. Claffy. Navigability of complex networks. *Nat Phys*, 5 (1):74 – 80, 2009. DOI: 10.1038/nphys1130 107

Angela Bonifati and Alfredo Cuzzocrea. Storing and retrieving xpath fragments in structured p2p networks. *Data & Knowl. Eng.*, 59(2):247 – 269, 2006. DOI: 10.1016/j.datak.2006.01.011 92

Nicolas Bruno, Nick Koudas, and Divesh Srivastava. Holistic twig joins: optimal xml pattern matching. In *Proc. ACM SIGMOD Int. Conf. on Management of Data*, pages 310–321, 2002. DOI: 10.1145/564691.564727 89

Benjamin Bustos, Gonzalo Navarro, and Edgar Chavez. Pivot selection techniques for proximity searching in metric spaces. *Pattern Recognition Letters*, 24(14):2357–2366, 2003. DOI: 10.1016/S0167-8655(03)00065-5 105

John Byers, Jeffrey Considine?, and Michael Mitzenmacher. Simple load balancing for distributed hash tables. In *Proc. 2nd Int. Workshop Peer-to-Peer Systems*, pages 80–87, 2003. DOI: 10.1007/978-3-540-45172-3_7 9

Min Cai and Martin Frank. Rdfpeers: a scalable distributed rdf repository based on a structured peer-to-peer network. In *Proc. 12th Int. World Wide Web Conf.*, pages 650–657, 2004. DOI: 10.1145/988672.988760 51

Min Cai and Kai Hwang. Distributed aggregation algorithms with load-balancing for scalable grid resource monitoring. In *Proc. 21st Int. Parallel & Distributed Processing Symp.*, pages 1–10, 2007. DOI: 10.1109/IPDPS.2007.370313 44

Min Cai, Martin R. Frank, Jinbo Chen, and Pedro A. Szekely. Maan: A multi-attribute addressable network for grid information services. *J. Grid Comp.*, 2(1):3–14, 2004. DOI: 10.1007/s10723-004-1184-y 21, 37

James P. Callan, Zhihong Lu, and W. Bruce Croft. Searching distributed collections with inference networks. In *Proc. 18th Annual Int. ACM SIGIR Conf. on Research and Development in Information Retrieval*, pages 21–28, 1995. DOI: 10.1145/215206.215328 86

Diego Calvanese, Giuseppe De Giacomo, Maurizio Lenzerini, and Riccardo Rosati. Logical foundations of peer-to-peer data integration. In *Proc. 23rd ACM SIGACT-SIGMOD-SIGART Symp. on Principles of Database Systems*, pages 241–251, 2004a. DOI: 10.1145/1055558.1055593 67

Diego Calvanese, Maurizio Lenzerini, Riccardo Rosati, and Guido Vetere. Hyper: A framework for peer-to-peer data integration on grids. In *Semantics for Grid Databases, First International IFIP Conference on Semantics of a Networked World*, volume 3226 of *Lecture Notes in Computer Science*, pages 144–157. Springer, 2004b. DOI: 10.1007/978-3-540-30145-5_9 67

Mitch Cherniack, Hari Balakrishnan, Magdalena Balazinska, Donald Carney, Ugur Çetintemel, Ying Xing, and Stanley B. Zdonik. Scalable distributed stream processing. In *Proc. 1st Biennial Conf. on Innovative Data Systems Research*, 2003. 110

Flavio Chierichetti, Ravi Kumar, and Prabhakar Raghavan. Compressed web indexes. In *Proc. 18th Int. World Wide Web Conf.*, pages 451–460, 2009. DOI: 10.1145/1526709.1526770 78

Stijn Christiaens. Metadata mechanisms: From ontology to folksonomy ... and back. In *On the Move to Meaningful Internet Systems 2006: OTM 2006 Workshops*, volume 4277 of *Lecture Notes in Computer Science*, pages 199–207. Springer Berlin / Heidelberg, 2006. DOI: 10.1007/11915034_43 96

Ian Clarke, Oskar Sandberg, Brandon Wiley, and Theodore Hong. Freenet: A distributed anonymous information storage and retrieval system. In *Designing Privacy Enhancing Technologies*, volume 2009 of *Lecture Notes in Computer Science*, pages 46–66. Springer Berlin / Heidelberg, 2001. DOI: 10.1007/3-540-44702-4_4 10, 111

Adina Crainiceanu, Prakash Linga, Johannes Gehrke, and Jayavel Shanmugasundaram. Querying peer-to-peer networks using p-trees. In *Proc. 7th Int. Workshop on the World Wide Web and Databases*, pages 25–30, New York, NY, USA, 2004. DOI: 10.1145/1017074.1017082 32

Adina Crainiceanu, Prakash Linga, Ashwin Machanavajjhala, Johannes Gehrke, and Jayavel Shanmugasundaram. P-ring: an efficient and robust p2p range index structure. In *Proc. ACM SIGMOD Int. Conf. on Management of Data*, pages 223–234, New York, NY, USA, 2007. DOI: 10.1145/1247480.1247507 30

Arturo Crespo and Hector Garcia-Molina. Routing indices for peer-to-peer systems. In *Proc. 22nd Int. Conf. on Distributed Computing Systems*, pages 23–33, 2002. DOI: 10.1109/ICDCS.2002.1022239 101

Arturo Crespo and Hector Garcia-Molina. Semantic overlay networks for p2p systems. In *Proc. 4th Int. Workshop Agents and Peer-to-Peer Computing*, pages 1–13, 2005. DOI: 10.1007/11574781_1 97

Bruno Crispo, Swaminathan Sivasubramanian, Pietro Mazzoleni, and Elisa Bertino. P-hera: Scalable fine-grained access control for p2p infrastructures. In *Proc. IEEE Int. Conf. on Parallel and Distributed Systems*, pages 585–591, 2005. DOI: 10.1109/ICPADS.2005.215 111

P. Cudre-Mauroux, K. Aberer, and A. Feher. Probabilistic message passing in peer data management systems. In *Proc. 22nd Int. Conf. on Data Engineering*, pages 41–52, 2006. DOI: 10.1109/MIC.2007.108 70, 72

Philippe Cudre-Mauroux, Suchit Agarwal, and Karl Aberer. Gridvine: An infrastructure for peer information management. *IEEE Internet Comput.*, 11(5):36–44, 2007. DOI: 10.1109/MIC.2007.108 51, 57, 58

Anwitaman Datta. Peer-to-peer storage. In Ling Liu and M. Tamer Özsu, editors, *Encyclopedia of Database Systems*, pages 2075–2081. Springer US, 2009. 109

Anwitaman Datta and Karl Aberer. Internet-scale storage systems under churn – a study of the steady-state using markov models. In *Proc. 6th IEEE Int. Conf. on Peer-to-Peer Computing*, pages 133–144, Los Alamitos, CA, USA, 2006. DOI: 10.1109/P2P.2006.22 9

Anwitaman Datta, Manfred Hauswirth, and Karl Aberer. Updates in highly unreliable, replicated peer-to-peer systems. In *Proc. 23rd Int. Conf. on Distributed Computing Systems*, pages 76–, 2003. DOI: 10.1109/ICDCS.2003.1203454 9

Anwitaman Datta, Manfred Hauswirth, Renault John, Roman Schmidt, and Karl Aberer. Range queries in trie-structured overlays. In *Proc. 5th IEEE Int. Conf. on Peer-to-Peer Computing*, pages 57–66, Los Alamitos, CA, USA, 2005. DOI: 10.1109/P2P.2005.31 34

Souptik Datta, Kanishka Bhaduri, Chris Giannella, Ran Wolff, and Hillol Kargupta. Distributed data mining in peer-to-peer networks. *IEEE Internet Comput.*, 10(4):18–26, 2006. DOI: 10.1109/MIC.2006.74 50, 109

Scott Deerwester, Susan T. Dumais, George W. Furnas, Thomas K. Landauer, and Richard Harshman. Indexing by latent semantic analysis. *J. American. Soc. for Information Sci.*, 41(6):391–407, 1990. DOI: 10.1002/(SICI)1097-4571(199009)41:6%3C391::AID-ASI1%3E3.0.CO;2-9 98

Alan Demers, Dan Greene, Carl Hauser, Wes Irish, John Larson, Scott Shenker, Howard Sturgis, Dan Swinehart, and Doug Terry. Epidemic algorithms for replicated database maintenance. In *Proc. ACM SIGACT-SIGOPS Symp. on Principles of Distributed Computing*, pages 1–12, 1987. DOI: 10.1145/41840.41841 45

Zoran Despotovic. Trust and reputation in peer-to-peer systems. In Ling Liu and M. Tamer Özsu, editors, *Encyclopedia of Database Systems*, pages 3183–3187. Springer US, 2009. 111

David DeWitt and Jim Gray. Parallel database systems: the future of high performance database systems. *Commun. ACM*, 35:85–98, 1992. DOI: 10.1145/129888.129894 20

Roger Dingledine, Michael Freedman, and David Molnar. The free haven project: Distributed anonymous storage service. In *Designing Privacy Enhancing Technologies*, volume 2009 of *Lecture Notes in Computer Science*, pages 67–95. Springer Berlin / Heidelberg, 2001. DOI: 10.1007/3-540-44702-4_5 111

Christos Doulkeridis, Kjetil Norvag, and Michalis Vazirgiannis. Desent: decentralized and distributed semantic overlay generation in p2p networks. *IEEE Journal on Selected Areas in Communications*, 25(1):25–34, 2007. DOI: 10.1109/JSAC.2007.070104 98, 102

Sameh El-Ansary, Luc Alima, Per Brand, and Seif Haridi. Efficient broadcast in structured p2p networks. In *Proc. 2nd Int. Workshop Peer-to-Peer Systems*, pages 304–314, 2003. DOI: 10.1007/978-3-540-45172-3_28 39

Ronald Fagin. Combining fuzzy information from multiple systems. *J. Comput. Syst. Sci.*, 58:83–99, 1999. DOI: 10.1006/jcss.1998.1600 85

Ronald Fagin, Phokion G. Kolaitis, Renée J. Miller, and Lucian Popa. Data exchange: Semantics and query answering. In *Proc. 9th Int. Conf. on Database Theory*, pages 207–224, 2003. DOI: 10.1007/3-540-36285-1_14 55, 67

Fabrizio Falchi, Claudio Gennaro, and Pavel Zezula. A content–addressable network for similarity search in metric spaces. In *Databases, Information Systems, and Peer-to-Peer Computing*, volume 4125 of *Lecture Notes in Computer Science*, pages 98–110. Springer Berlin / Heidelberg, 2007. DOI: 10.1007/978-3-540-71661-7_9 98, 105

Pascal Felber and Ernst W. Biersack. Peer-to-peer content distribution. In Ling Liu and M. Tamer Özsu, editors, *Encyclopedia of Database Systems*, pages 2061–2065. Springer US, 2009. 110

Philippe Flajolet and G. Nigel Martin. Probabilistic counting. In *Proc. 24th Annual Symp. on Foundations of Computer Science*, pages 76–82, 1983. 43

Leonidas Galanis, Yuan Wang, Shawn R. Jeffery, and David J. DeWitt. Locating data sources in large distributed systems. In *Proc. 29th Int. Conf. on Very Large Data Bases*, pages 874–885, 2003. 89, 90, 93

Herve Gallaire and Jack Minker. *Logic and Data Bases*. Perseus Publishing, 1978. 61

Wojciech Galuba and Karl Aberer. Generic emergent overlays in arbitrary peer identifier spaces. In *Self-Organizing Systems, Second International Workshop*, volume 4725 of *Lecture Notes in Computer Science*, pages 88–102. Springer, 2007. DOI: 10.1007/978-3-540-74917-2_9 10

Wojtek Galuba, Karl Aberer, Zoran Despotovic, and Wolfgang Kellerer. Leveraging social networks for increased bittorrent robustness. In *7th IEEE Consumer Communications and Networking Conference*, pages 1–5, 2010. DOI: 10.1109/CCNC.2010.5421823 111

Prasanna Ganesan, Mayank Bawa, and Hector Garcia-Molina. Online balancing of range-partitioned data with applications to peer-to-peer systems. In *Proc. 30th Int. Conf. on Very Large Data Bases*, pages 444–455, 2004a. 24, 25

Prasanna Ganesan, Beverly Yang, and Hector Garcia-Molina. One torus to rule them all: multi-dimensional queries in p2p systems. In *Proc. 7th Int. Workshop on the World Wide Web and Databases*, pages 19–24, New York, NY, USA, 2004b. DOI: 10.1145/1017074.1017081 37, 38

Luis Garcés-Erice, Pascal Felber, Ernst W. Biersack, Guillaume Urvoy-Keller, and Keith W. Ross. Data indexing in peer-to-peer dht networks. In *Proc. 23rd Int. Conf. on Distributed Computing Systems*, pages 200–208, 2004. DOI: 10.1109/ICDCS.2004.1281584 91

Aristides Gionis, Piotr Indyk, and Rajeev Motwani. Similarity search in high dimensions via hashing. In *Proc. 25th Int. Conf. on Very Large Data Bases*, pages 518–529, 1999. 105

Sarunas Girdzijauskas, Anwitaman Datta, and Karl Aberer. On small world graphs in non-uniformly distributed key spaces. In *21st Int. Conf. on Data Engineering Workshops*, pages 1187–, 2005. DOI: 10.1109/ICDE.2005.254 10

Šarūnas Girdzijauskas, Anwitaman Datta, and Karl Aberer. Structured overlay for heterogeneous environments: Design and evaluation of oscar. *ACM Trans. Auton. Adapt. Syst.*, 5(1):1–25, 2010. DOI: 10.1145/1671948.1671950 10, 29

Christos Gkantsidis, Milena Mihail, and Amin Saberi. Random walks in peer-to-peer networks: Algorithms and evaluation. *Performance Evaluation*, 63(3):241–263, 2006. DOI: 10.1016/j.peva.2005.01.002 46

P. Krishna Gummadi, Ramakrishna Gummadi, Steven D. Gribble, Sylvia Ratnasamy, Scott Shenker, and Ion Stoica. The impact of dht routing geometry on resilience and proximity. In *Proc. ACM Int. Conf. on Data Communication*, pages 381–394, New York, NY, USA, 2003. DOI: 10.1145/863955.863998 9

Peter J. Haas and Joseph M. Hellerstein. Ripple joins for online aggregation. In *Proc. ACM SIGMOD Int. Conf. on Management of Data*, pages 287–298, 1999. DOI: 10.1145/304181.304208 47, 48

Peter Haase, Ronny Siebes, and Frank van Harmelen. Peer selection in peer-to-peer networks with semantic topologies. In *Semantics for Grid Databases, First International IFIP Conference on Semantics of a Networked World*, volume 3226 of *Lecture Notes in Computer Science*, pages 108–125. Springer Berlin / Heidelberg, 2004. DOI: 10.1007/978-3-540-30145-5_7 97, 98, 102

Parisa Haghani, Sebastian Michel, and Karl Aberer. Distributed similarity search in high dimensions using locality sensitive hashing. In *Advances in Database Technology, Proc. 12th Int. Conf. on Extending Database Technology*, pages 744–755, 2009. DOI: 10.1145/1516360.1516446 98, 105, 106, 109

Alon Y. Halevy, Zachary G. Ives, Dan Suciu, and Igor Tatarinov. Schema mediation in peer data management systems. In *Proc. 19th Int. Conf. on Data Engineering*, pages 505–, 2003. DOI: 10.1109/ICDE.2003.1260817 63, 64, 66, 67

Andreas Harth and Stefan Decker. Optimized index structures for querying rdf from the web. In *Third Latin American Web Congress*, pages 71–80, 2005. DOI: 10.1109/LAWEB.2005.25 52

Nicholas J. A. Harvey, Michael B. Jones, Stefan Saroiu, Marvin Theimer, and Alec Wolman. Skipnet: a scalable overlay network with practical locality properties. In *Proc. 4th USENIX Symp. on Internet Tech. and Systems*, pages 9–9, Berkeley, CA, USA, 2003. 30, 31

Joseph M. Hellerstein. Toward network data independence. *ACM SIGMOD Rec.*, 32:34–40, 2003. DOI: 10.1145/945721.945730 14

Joseph M. Hellerstein, Peter J. Haas, and Helen J. Wang. Online aggregation. In *Proc. ACM SIGMOD Int. Conf. on Management of Data*, pages 171–182, 1997. DOI: 10.1145/253262.253291 47, 48

Ryan Huebsch, Joseph M. Hellerstein, Nick Lanham, Boon Thau Loo, Scott Shenker, and Ion Stoica. Querying the internet with pier. In *Proc. 29th Int. Conf. on Very Large Data Bases*, pages 321–332, 2003. 38, 40

Zachary G. Ives. Updates and transactions in peer-to-peer systems. In Ling Liu and M. Tamer Özsu, editors, *Encyclopedia of Database Systems*, pages 3241–3244. Springer US, 2009. 110

H. V. Jagadish, Beng Chin Ooi, and Quang Hieu Vu. Baton: a balanced tree structure for peer-to-peer networks. In *Proc. 31st Int. Conf. on Very Large Data Bases*, pages 661–672, 2005. 31

Mark Jelasity and Alberto Montresor. Epidemic-style proactive aggregation in large overlay networks. In *Proc. 23rd Int. Conf. on Distributed Computing Systems*, pages 102–109, Washington, DC, USA, 2004. DOI: 10.1109/ICDCS.2004.1281573 45

Audun Josang, Roslan Ismail, and Colin Boyd. A survey of trust and reputation systems for online service provision. *Decision Support Systems*, 43(2):618 – 644, 2007. DOI: 10.1016/j.dss.2005.05.019 111

Verena Kantere, Dimitrios Tsoumakos, Timos Sellis, and Nick Roussopoulos. Groupeer: Dynamic clustering of p2p databases. *Information Systems*, 34(1):62–86, 2009. DOI: 10.1016/j.is.2008.04.002 68

David R. Karger and Matthias Ruhl. Simple efficient load-balancing algorithms for peer-to-peer systems. *Theory Comput. Syst.*, 39:787–804, 2006. DOI: 10.1007/s00224-006-1246-6 25

Anastasios Kementsietsidis and Marcelo Arenas. Data sharing through query translation in autonomous sources. In *Proc. 30th Int. Conf. on Very Large Data Bases*, pages 468–479. VLDB Endowment, 2004. 58, 59, 60

Anastasios Kementsietsidis, Marcelo Arenas, and Renée J. Miller. Mapping data in peer-to-peer systems: semantics and algorithmic issues. In *Proc. ACM SIGMOD Int. Conf. on Management of Data*, pages 325–336, New York, NY, USA, 2003. DOI: 10.1145/872757.872798 59

Jon M. Kleinberg. The small-world phenomenon: an algorithm perspective. In *Proc. 32nd Annual ACM Symp. on Theory of Computing*, pages 163–170, 2000. DOI: 10.1145/335305.335325 10

Fabius Klemm, Jean-Yves Le Boudec, and Karl Aberer. Congestion control for distributed hash tables. In *Proc. IEEE Int. Symp. Network Computing and Applications*, pages 189–195, 2006. DOI: 10.1109/NCA.2006.19 87

Fabius Klemm, Sarunas Girdzijauskas, Jean-Yves Le Boudec, and Karl Aberer. On routing in distributed hash tables. In *Proc. 7th IEEE Int. Conf. on Peer-to-Peer Computing*, pages 113–122, Los Alamitos, CA, USA, 2007. DOI: 10.1109/P2P.2007.38 30

Christoph Koch. Query rewriting with symmetric constraints. *AI Commun.*, 17:41–55, 2004. 63, 67

Georgia Koloniari and Evaggelia Pitoura. Content-based routing of path queries in peer-to-peer systems. In *Advances in Database Technology, Proc. 9th Int. Conf. on Extending Database Technology*, pages 525–526, 2004. 93, 94

Georgia Koloniari, Yannis Petrakis, and Evaggelia Pitoura. Content-based overlay networks for xml peers based on multi-level bloom filters. In *Databases, Information Systems, and Peer-to-Peer Computing*, volume 2944 of *Lecture Notes in Computer Science*, pages 232–247. Springer Berlin / Heidelberg, 2004. DOI: 10.1007/978-3-540-24629-9_17 93

Hanna Köpcke and Erhard Rahm. Frameworks for entity matching: A comparison. *Data & Knowl. Eng.*, 69(2):197–210, 2010. DOI: 10.1016/j.datak.2009.10.003 58

Dmitry Korzun and Andrei Gurtov. Survey on hierarchical routing schemes in flatdistributed hash tables. *Peer-to-Peer Networking and Applications*, pages 1–30, 2010. DOI: 10.1007/s12083-010-0093-z 1, 6, 7

Vibhore Kumar, Brian F. Cooper, Zhongtang Cai, Greg Eisenhauer, and Karsten Schwan. Resource-aware distributed stream management using dynamic overlays. In *Proc. 23rd Int. Conf. on Distributed Computing Systems*, pages 783–792, 2005. DOI: 10.1109/ICDCS.2005.69 110

Richard Kuntschke, Bernhard Stegmaier, Alfons Kemper, and Angelika Reiser. Streamglobe: Processing and sharing data streams in grid-based p2p infrastructures. In *Proc. 31st Int. Conf. on Very Large Data Bases*, pages 1259–1262, 2005. 110

Maurizio Lenzerini. Data integration: a theoretical perspective. In *Proc. ACM SIGMOD Int. Conf. on Management of Data*, pages 233–246, 2002. DOI: 10.1145/543613.543644 61, 63

Dongsheng Li, Jiannong Cao, Xicheng Lu, and Kaixian Chen. Efficient range query processing in peer-to-peer systems. *IEEE Trans. Knowl. and Data Eng.*, 21(1):78–91, 2009. DOI: 10.1109/TKDE.2008.99 21

Guoliang Li, Beng Ooi, Bei Yu, and Lizhu Zhou. Schema mapping in p2p networks based on classification and probing. In Ramamohanarao Kotagiri, P. Krishna, Mukesh Mohania, and Ekawit Nantajeewarawat, editors, *Proc. 12th Int. Conf. on Database Systems for Advanced Applications*, pages 688–702, 2007. 69

Ji Li, Karen Sollins, and Dah-Yoh Lim. Implementing aggregation and broadcast over distributed hash tables. *SIGCOMM Comput. Commun. Rev.*, 35:81–92, 2005. DOI: 10.1145/1052812.1052813 44

Jin Li. On peer-to-peer (p2p) content delivery. *Peer-to-Peer Networking and Applications*, 1:45–63, 2008. DOI: 10.1007/s12083-007-0003-1 110

Jinyang Li, Boon Loo, Joseph Hellerstein, M. Kaashoek, David Karger, and Robert Morris. On the feasibility of peer-to-peer web indexing and search. In *Proc. 2nd Int. Workshop Peer-to-Peer Systems*, pages 207–215, 2003. DOI: 10.1007/978-3-540-45172-3_19 76, 80, 81

Mei Li, Wang-Chien Lee, and A. Sivasubramaniam. Semantic small world: an overlay network for peer-to-peer search. In *Proc. of the 12th IEEE Int. Conf. on Network Protocols*, pages 228–238, 2004. DOI: 10.1109/ICNP.2004.1348113 98, 106

Erietta Liarou, Stratos Idreos, and Manolis Koubarakis. Evaluating conjunctive triple pattern queries over large structured overlay networks. In *Proc. 5th Int. Semantic Web Conf.*, pages 399–413, 2006. DOI: 10.1007/11926078_29 51, 52, 53

Alexander Löser, Wolf Siberski, Martin Wolpers, and Wolfgang Nejdl. Information integration in schema-based peer-to-peer networks. In *Proc. 15th Int. Conf. on Advanced Information Systems Eng.*, pages 258–272, 2003. 57

Alexander Löser, Christoph Tempich, Bastian Quilitz, Wolf-Tilo Balke, Steffen Staab, and Wolfgang Nejdl. Searching dynamic communities with personal indexes. In *Proc. 4th Int. Semantic Web Conf.*, pages 491–505, 2005. DOI: 10.1007/11574620_36 101

Eng Keong Lua, Jon Crowcroft, Marcelo Pias, Ravi Sharma, and Steven Lim. A survey and comparison of peer-to-peer overlay network schemes. *IEEE Communications Surveys and Tutorials*, 7 (2):72 – 93, 2005. DOI: 10.1109/COMST.2005.1610546 1

Robert W.P. Luk, H.V. Leong, Tharam S. Dillon, Alvin T.S. Chan, W. Bruce Croft, and James Allan. A survey in indexing and searching xml documents. *J. American. Soc. for Information Sci. & Tech.*, 53(6):415–437, 2002. DOI: 10.1002/asi.10056 88

Qin Lv, Pei Cao, Edith Cohen, Kai Li, and Scott Shenker. Search and replication in unstructured peer-to-peer networks. In *Proc. 16th Annual Int. Conf. on Supercomputing*, pages 84–95, 2002. DOI: 10.1145/514191.514206 5, 9

Gurmeet Singh Manku, Mayank Bawa, and Prabhakar Raghavan. Symphony: distributed hashing in a small world. In *Proc. 4th USENIX Symp. on Internet Tech. and Systems*, USITS'03, pages 10–. USENIX Association, 2003. 10

Mehedi Masud and Iluju Kiringa. Transaction processing in a peer to peer database network. *Data & Knowl. Eng.*, 70(4):307–334, 2011. DOI: 10.1016/j.datak.2010.12.003 110

Petar Maymounkov and David Mazières. Kademlia: A peer-to-peer information system based on the xor metric. In *Proc. 1st Int. Workshop Peer-to-Peer Systems*, pages 53–65, 2002. DOI: 10.1007/3-540-45748-8_5 6, 7

Elke Michlmayr, Arno Pany, and Gerti Kappel. Using taxonomies for content-based routing with ants. *Comp. Netw.*, 51(16):4514–4528, 2007. DOI: 10.1016/j.comnet.2007.06.015 97, 101

George A. Miller. Wordnet: a lexical database for english. *Commun. ACM*, 38:39–41, 1995. DOI: 10.1145/219717.219748 96

Dejan S. Milojicic, Vana Kalogeraki, Rajan Lukose, Kiran Nagaraja1, Jim Pruyne, Bruno Richard, Sami Rollins, and Zhichen Xu. Peer-to-peer computing. Technical Report HPL-2002-57, HP Laboratories Palo Alto, 2002. 1

Wolfgang Nejdl, Martin Wolpers, Wolf Siberski, Christoph Schmitz, Mario Schlosser, Ingo Brunkhorst, and Alexander Löser. Super-peer-based routing strategies for rdf-based peer-to-peer networks. In *Proc. 12th Int. World Wide Web Conf.*, pages 177–186, 2004. DOI: 10.1016/j.websem.2003.11.004 53

Wee Siong Ng, Beng Chin Ooi, Kian-Lee Tan, and Aoying Zhou. Peerdb: a p2p-based system for distributed data sharing. In *Proc. 19th Int. Conf. on Data Engineering*, pages 633–644, 2003. DOI: 10.1109/ICDE.2003.1260827 69

Andy Oram. *Peer-to-Peer: Harnessing the Power of Disruptive Technologies.* O'Reilly & Associates, Inc., Sebastopol, CA, USA, 2001. 1

Jack A. Orenstein and T. H. Merrett. A class of data structures for associative searching. In *Proc. 3rd ACM SIGACT-SIGMOD Symp. on Principles of Database Systems*, pages 181–190, New York, NY, USA, 1984. 37

Josiane Xavier Parreira, Sebastian Michel, and Gerhard Weikum. p2pdating: Real life inspired semantic overlay networks for web search. *Information Processing & Management*, 43(3):643–664, 2007. Special Issue on Heterogeneous and Distributed IR. DOI: 10.1016/j.ipm.2006.09.007 98, 102

Wilma Penzo, Stefano Lodi, Federica Mandreoli, Riccardo Martoglia, and Simona Sassatelli. Semantic peer, here are the neighbors you want! In *Advances in Database Technology, Proc. 11th Int. Conf. on Extending Database Technology*, pages 26–37, 2008. DOI: 10.1145/1353343.1353351 97, 98, 102

Rachel Pottinger and Alon Halevy. Minicon: A scalable algorithm for answering queries using views. *VLDB J.*, 10:182–198, 2001. DOI: 10.1007/s007780100048 67

Johan A. Pouwelse, Pawel Garbacki, Jun Wang, Arno Bakker, Jie Yang, Alexandru Iosup, Dick H. J. Epema, Marcel J. T. Reinders, Maarten van Steen, and Henk J. Sips. Tribler: a social-based peer-to-peer system. *Concurrency and Computation: Practice and Experience*, 20(2):127–138, 2008. DOI: 10.1002/cpe.1189 111

Paraskevi Raftopoulou and Euripides G. M. Petrakis. icluster: a self-organizing overlay network for p2p information retrieval. In *Proc. 30th European Conf. on IR Research*, pages 65–76, 2008. DOI: 10.1007/978-3-540-78646-7_9 98, 99, 102

Erhard Rahm and Philip A. Bernstein. A survey of approaches to automatic schema matching. *VLDB J.*, 10:334–350, 2001. DOI: 10.1007/s007780100057 55, 67

Ananth Rao, Karthik Lakshminarayanan, Sonesh Surana, Richard Karp, and Ion Stoica. Load balancing in structured p2p systems. In *Proc. 2nd Int. Workshop Peer-to-Peer Systems*, pages 68–79, 2003. DOI: 10.1007/978-3-540-45172-3_6 9

Praveen Rao and Bongki Moon. Locating xml documents in a peer-to-peer network using distributed hash tables. *IEEE Trans. Knowl. and Data Eng.*, 21(12):1737–1752, 2009. DOI: 10.1109/TKDE.2009.26 94

Amir H. Rasti, Daniel Stutzbach, and Reza Rejaie. On the long-term evolution of the two-tier gnutella overlay. In *Proc. 25th Annual Joint Conf. of the IEEE Computer and Communication Societies*, pages 1–6, 2006. DOI: 10.1109/INFOCOM.2006.44 10

Sylvia Ratnasamy, Paul Francis, Mark Handley, Richard Karp, and Scott Shenker. A scalable content-addressable network. In *Proc. ACM Int. Conf. on Data Communication*, pages 161–172, New York, NY, USA, 2001. DOI: 10.1145/964723.383072 6, 9, 38

Patrick Reynolds and Amin Vahdat. Efficient peer-to-peer keyword searching. In *Proc. ACM/IFIP/USENIX Int. Middleware Conf.*, pages 21–40, 2003. DOI: 10.1007/3-540-44892-6_2 80, 81, 83

Matei Ripeanu. Peer-to-peer architecture case study: Gnutella network. In *Proc. 1st IEEE Int. Conf. on Peer-to-Peer Computing*, pages 99–100, 2001. DOI: 10.1109/P2P.2001.990433 5

John Risson and Tim Moors. Survey of research towards robust peer-to-peer networks: Search methods. *Comp. Netw.*, 50(17):3485–3521, 2006. DOI: 10.1016/j.comnet.2006.02.001 1

Antony I. T. Rowstron and Peter Druschel. Pastry: Scalable, decentralized object location, and routing for large-scale peer-to-peer systems. In *Proc. IFIP/ACM Int. Conf. on Distributed Systems Platforms*, pages 329–350, 2001. DOI: 10.1007/3-540-45518-3_18 7, 10

Ozgur D. Sahin, Abhishek Gupta, Divyakant Agrawal, and Amr El Abbadi. A peer-to-peer framework for caching range queries. In *Proc. 20th Int. Conf. on Data Engineering*, pages 165–176, Los Alamitos, CA, USA, 2004. DOI: 10.1109/ICDE.2004.1319993 35

Ozgur D. Sahin, Fatih Emekci, Divyakant Agrawal, and Amr El Abbadi. Content-based similarity search over peer-to-peer systems. In *Databases, Information Systems, and Peer-to-Peer Computing*, volume 3367 of *Lecture Notes in Computer Science*, pages 61–78. Springer Berlin / Heidelberg, 2005. 106

Nima Sarshar, P. Oscar Boykin, and Vwani P. Roychowdhury. Percolation search in power law networks: Making unstructured peer-to-peer networks scalable. In *Proc. 4th IEEE Int. Conf. on Peer-to-Peer Computing*, pages 2–9, 2004. DOI: 10.1109/PTP.2004.1334925 5

Carlo Sartiani, Paolo Manghi, Giorgio Ghelli, and Giovanni Conforti. Xpeer : A self-organizing xml p2p database system. In *Current Trends in Database Technology - EDBT 2004 Workshops*, volume 3268 of *Lecture Notes in Computer Science*, pages 429–432. Springer Berlin / Heidelberg, 2005. DOI: 10.1007/978-3-540-30192-9_45 94

Christoph Schmitz. Self-organization of a small world by topic. In *LWA 2004: Lernen - Wissensentdeckung - Adaptivität*, pages 303–310, 2004. 97, 102

Rüdiger Schollmeier, Ingo Gruber, and Michael Finkenzeller. Routing in mobile ad-hoc and peer-to-peer networks a comparison. In *NETWORKING Workshops*, volume 2376 of *Lecture Notes in Computer Science*, pages 172–186. Springer, 2002. DOI: 10.1007/3-540-45745-3_16 110

Thorsten Schuett, Florian Schintke, and Alexander Reinefeld. Range queries on structured overlay networks. *Computer Communications*, 31(2):280–291, 2008. Special Issue: Foundation of Peer-to-Peer Computing. DOI: 10.1016/j.comcom.2007.08.027 30, 38, 39

Giovanna Di Marzo Serugendo, Noria Foukia, Salima Hassas, Anthony Karageorgos, Soraya Kouadri Mostéfaoui, Omer F. Rana, Mihaela Ulieru, Paul Valckenaers, and Chris Van Aart. Self-organisation: Paradigms and applications. In Giovanna Di Marzo Serugendo, Anthony Karageorgos, Omer F. Rana, and Franco Zambonelli, editors, *Engineering Self-Organising Systems*, volume 2977 of *Lecture Notes in Computer Science*, pages 1–19. Springer Berlin / Heidelberg, 2004. 2

Yanfeng Shu, Beng Chin Ooi, Kian-Lee Tan, and Aoying Zhou. Supporting multi-dimensional range queries in peer-to-peer systems. In *Proc. 5th IEEE Int. Conf. on Peer-to-Peer Computing*, pages 173–180, Los Alamitos, CA, USA, 2005. IEEE Computer Society. DOI: 10.1109/P2P.2005.35 37

Ronny Siebes and Spyros Kotoulas. proute: Peer selection using shared term similarity matrices. *Web Intelli. and Agent Systems*, 5:89–107, 2007. 98, 102

Gleb Skobeltsyn, Manfred Hauswirth, and Karl Aberer. Efficient processing of xpath queries with structured overlay networks. In *Proc. Confederated Int. Conf. DOA, CoopIS and ODBASE*, pages 1243–1260, 2005. 90, 91

Gleb Skobeltsyn, Toan Luu, Ivana Podnar Zarko, Martin Rajman, and Karl Aberer. Web text retrieval with a p2p query-driven index. In *Proc. 33rd Annual Int. ACM SIGIR Conf. on Research and Development in Information Retrieval*, pages 679–686, 2007. DOI: 10.1145/1277741.1277857 82, 83

Kunwadee Sripanidkulchai, Bruce Maggs, and Hui Zhang. Efficient content location using interest-based locality in peer-to-peer systems. In *Proc. 22nd Annual Joint Conf. of the IEEE Computer and Communication Societies*, pages 2166–2176, 2003. DOI: 10.1109/INFCOM.2003.1209237 100

Ion Stoica, Robert Morris, David Karger, M. Frans Kaashoek, and Hari Balakrishnan. Chord: A scalable peer-to-peer lookup service for internet applications. In *Proc. ACM Int. Conf. on Data Communication*, pages 149–160, 2001. DOI: 10.1145/383059.383071 6, 7

Torsten Suel, Chandan Mathur, Jo wen Wu, Jiangong Zhang, Alex Delis, Mehdi Kharrazi, Xiaohui Long, and Kulesh Shanmugasundaram. Odissea: A peer-to-peer architecture for scalable web search and information retrieval. In *Proc. 6th Int. Workshop on the World Wide Web and Databases*, pages 67–72, 2003. 84

Chunqiang Tang, Zhichen Xu, and Sandhya Dwarkadas. Peer-to-peer information retrieval using self-organizing semantic overlay networks. In *Proc. ACM Int. Conf. on Data Communication*, pages 175–186, 2003. DOI: 10.1145/863955.863976 98, 104

Christoph Tempich, Steffen Staab, and Adrian Wranik. Remindin': semantic query routing in peer-to-peer networks based on social metaphors. In *Proc. 12th Int. World Wide Web Conf.*, pages 640–649, 2004. DOI: 10.1145/988672.988759 101

Juan M. Tirado, Daniel Higuero, Florin Isaila, Jesus Carretero, and Adriana Iamnitchi. Affinity p2p: A self-organizing content-based locality-aware collaborative peer-to-peer network. *Comp. Netw.*, 54(12):2056–2070, 2010. DOI: 10.1016/j.comnet.2010.04.016 97, 103

Peter Triantafillou and Ioannis Aekaterinidis. Peer-to-peer publish-subscribe systems. In Ling Liu and M. Tamer Özsu, editors, *Encyclopedia of Database Systems*, pages 2069–2075. Springer US, 2009. 110

Dimitrios Tsoumakos and Nick Roussopoulos. Analysis and comparison of p2p search methods. In *Proc. of the 1st Int. Conf. on Scalable information systems*, 2006. DOI: 10.1145/1146847.1146872 5

Quang Hieu Vu, Beng Chin Ooi, Martin Rinard, and Kian-Lee Tan. Histogram-based global load balancing in structured peer-to-peer systems. *IEEE Trans. Knowl. and Data Eng.*, 21(4):595–608, 2009. DOI: 10.1109/TKDE.2008.182 26

Quang Hieu Vu, Mihai Lupu, and Beng Chin Ooi. *Peer-to-Peer Computing - Principles and Applications*. Springer, 2010. 1

Dan Wallach. A survey of peer-to-peer security issues. In *Software Security Ñ Theories and Systems*, volume 2609 of *Lecture Notes in Computer Science*, pages 253–258. Springer Berlin / Heidelberg, 2003. DOI: 10.1007/3-540-36532-X_4 111

Shiyuan Wang, Beng Chin Ooi, Anthony K. H. Tung, and Lizhen Xu. Efficient skyline query processing on peer-to-peer networks. In *Proc. 23rd Int. Conf. on Data Engineering*, pages 1126–1135, 2007. 109

Hakim Weatherspoon and John Kubiatowicz. Erasure coding vs. replication: A quantitative comparison. In *Proc. 1st Int. Workshop Peer-to-Peer Systems*, pages 328–337, 2002. DOI: 10.1007/3-540-45748-8_31 9

Roger Weber, Hans-Jörg Schek, and Stephen Blott. A quantitative analysis and performance study for similarity-search methods in high-dimensional spaces. In *Proc. 24th Int. Conf. on Very Large Data Bases*, pages 194–205, 1998. 98

Judith Winter and Oswald Drobnik. Spirix: A peer-to-peer search engine for xml-retrieval. In *Advances in Focused Retrieval*, volume 5631 of *Lecture Notes in Computer Science*, pages 97–105. Springer Berlin / Heidelberg, 2009. DOI: 10.1007/978-3-642-03761-0_11 92

Sai Wu, Shouxu Jiang, Beng Chin Ooi, and Kian-Lee Tan. Distributed online aggregations. In *Proc. 35th Int. Conf. on Very Large Data Bases*, pages 443–454, 2009. 47, 48

Zhichen Xu and Zheng Zhang. Building low-maintenance expressways for p2p systems. Technical Report HPL-2002-41, HP Laboratories Palo Alto, 2002. 9

Cheng Xiang Zhai. Statistical language models for information retrieval: A critical review. *Found. Trends Inf. Retr.*, 2:137–213, March 2008. DOI: 10.1561/1500000008 98

Ben Y. Zhao, Ling Huang, Jeremy Stribling, Sean C. Rhea, Anthony D. Joseph, and John Kubiatowicz. Tapestry: a resilient global-scale overlay for service deployment. *IEEE Journal on Selected Areas in Communications*, 22(1):41–53, 2004. DOI: 10.1109/JSAC.2003.818784 7

Justin Zobel and Alistair Moffat. Inverted files for text search engines. *ACM Comput. Surv.*, 38, 2006. DOI: 10.1145/1132956.1132959 76, 77

Author's Biography

KARL ABERER

Karl Aberer is a full professor for Distributed Information Systems at EPFL Lausanne, Switzerland, since 2000. Since 2005 he is the director of the Swiss National Research Center for Mobile Information and Communication Systems (NCCR-MICS, www.mics.ch). Prior to his current position, he was senior researcher at the Integrated Publication and Information Systems institute (IPSI) of GMD in Germany. He received his Ph.D. in mathematics in 1991 from the ETH Zürich. His research interests are on semantics and self-organization in information systems with applications in peer-to-peer search, semantic web, trust management and mobile and sensor networks.

He is or has been serving on the editorial boards of SIGMOD Record, VLDB Journal, ACM Transaction on Autonomous and Adaptive Systems and World Wide Web Journal and been co-chairing among others the ICDE, ISWC, MDM, ODBASE, P2P, VLDB and WISE conferences.

Index

CPSIA information can be obtained at www.ICGtesting.com
Printed in the USA
269481BV00003B/179-206/P

9 781608 457199